Animal Physiology and Morphology

Animal Physiology and Morphology

Samantha Morales

SYRAWOOD
PUBLISHING HOUSE

New York

Published by Syrawood Publishing House,
750 Third Avenue, 9th Floor,
New York, NY 10017, USA
www.syrawoodpublishinghouse.com

Animal Physiology and Morphology
Samantha Morales

© 2020 Syrawood Publishing House

International Standard Book Number: 978-1-64740-032-3 (Hardback)

Cataloging-in-Publication Data

Animal physiology and morphology / Samantha Morales.
    p. cm.
Includes bibliographical references and index.
ISBN 978-1-64740-032-3
1. Physiology. 2. Morphology (Animals). 3. Physiology, Comparative.
4. Morphology. I. Morales, Samantha.
QP31.2 .A55 2020
612--dc23

Table of Contents

Preface

This book has been written, keeping in view that students want more practical information. Thus, my aim has been to make it as comprehensive as possible for the readers. I would like to extend my thanks to my family and co-workers for their knowledge, support and encouragement all along.

The study of the life supporting processes in animals is known as animal physiology. It examines the regulation, integration and functioning of biological processes under various environmental conditions. The study of animal physiology is closely linked with anatomy. The study of the size, structure and shape of animals falls under the field of morphology. It is the study of the biological form and arrangement of the organs of the animals. It generally focuses on the bones, muscles, nerves and blood vessels which constitute the bodies of animals. The topics included in this book on animal physiology and morphology are of utmost significance and bound to provide incredible insights to readers. Those in search of information to further their knowledge will be greatly assisted by this book. Coherent flow of topics, student-friendly language and extensive use of examples make it an invaluable source of knowledge.

A brief description of the chapters is provided below for further understanding:

Chapter – Introduction to Physiology and Morphology

The branch of science which studies the life supporting properties, processes and functions of animals or their parts is termed as animal physiology. The study of the size, structure and shape of animals as well as their constituent parts is undertaken within the discipline of animal morphology. All the diverse aspects of animal morphology and physiology have been briefly introduced in this chapter.

Chapter – Vertebrate Physiology

The branch of biology which studies the functions and mechanisms inside vertebrates is termed as vertebrate physiology. Some of the important types of vertebrates studied within this field are birds and fish. This chapter closely examines the key concepts related to the physiology of these vertebrates to provide an extensive understanding of the subject.

Chapter – Invertebrate Physiology

The branch of biology which studies the physiology of animals which neither possess nor develop a vertebral column derived from the notochord is termed as invertebrate physiology. Some of the common invertebrates are insects and worms. This chapter discusses the diverse aspects related to these invertebrates in detail.

Chapter – Animal Cell Physiology

The biological branch which focuses on the activities which take place within cells is termed as cell physiology. Some of the processes studied within this field are cell adhesion, cell division and cell signaling. The topics elaborated in this chapter will help in gaining a better perspective about these areas of study within cell physiology.

Chapter – Animal Behavior

Animal behavior, also known as ethology, is a scientific field which studies the behavior of animals under natural conditions. It considers behavior as an evolutionary adaptive trait. The diverse focus areas within the field of animal behavior such as neuroethology and phenotypic plasticity have been thoroughly discussed in this chapter.

Samantha Morales

1
Introduction to Physiology and Morphology

The branch of science which studies the life supporting properties, processes and functions of animals or their parts is termed as animal physiology. The study of the size, structure and shape of animals as well as their constituent parts is undertaken within the discipline of animal morphology. All the diverse aspects of animal morphology and physiology have been briefly introduced in this chapter.

Physiology

Physiology can refer either to the parts or functions (mechanical, physical, and biochemical) of living organisms, or to the branch of biology that deals with the study of all the parts of living organisms and their various functions.

Since the function of a part is related to its structure, physiology naturally is related to anatomy, a term that can refer either to the internal structure and organization of an organism or any of its parts, or to the branch of biology that studies the internal structure and organization of living things.

Since the dawn of civilization, human beings have had a curiosity about nature and about the human body. In their efforts to better understand the mysteries of life, one key area is physiology. Most fields of biological endeavor—botany, zoology, embryology, cytology, etc.—include a study of function and thus of physiology. The science of medicine is particularly tied to the study of human physiology.

Physiology has traditionally been divided into plant physiology and animal physiology, but the principles of physiology are universal, no matter what particular organism is being studied. For example, what is learned about the physiology of yeast cells can also apply to human cells.

The field of animal physiology extends the tools and methods of human physiology to non-human animal species. Plant physiology borrows techniques from both fields. Physiology's scope of subjects is at least as diverse as the tree of life itself. Due to this diversity of subjects, research in animal physiology tends to concentrate on understanding how physiological traits changed throughout the history of animals.

Other major branches of scientific study with roots grounded in physiology research include biochemistry, biophysics, paleobiology, biomechanics, and pharmacology.

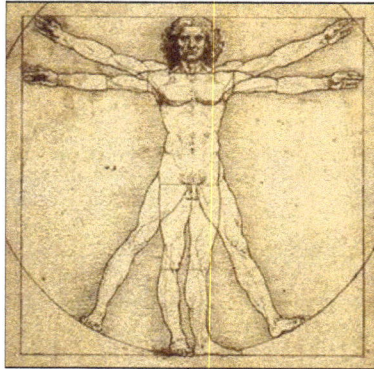

Leonardo da Vinci's Vitruvian Man, an important early achievement in the study of physiology.

Areas of Physiology

Human and Animal

Human physiology is the most complex area in physiology. This area has several subdivisions that overlap with each other. Many animals have similar anatomy to humans and share many of these areas:

- Myophysiology deals with the operation of muscles.

- Neurophysiology concerns the physiology of brains and nerves.

- Cell physiology addresses the functioning of individual cells.

- Comparative or Environmental physiology examines how animals adapt to their environment.

- Membrane physiology focuses on the exchange of molecules across the cell membrane.

- Respiratory physiology describes the mechanics of gas exchange at the cellular level and also at a gross anatomic level within the lungs.

- Circulation also known as cardiovascular physiology, deals with the heart, blood, and blood vessels and issues arising from any malfunction.

- Renal physiology focuses on the excretion of ions and other metabolites at the kidney.

- Endocrinology covers endocrine hormones which affect every cell in the body.

- Neuroendocrinology concerns the complex interactions of the neurological and endocrinological systems which together regulate physiology.

- Reproductive physiology concerns the reproductive cycle.

- Exercise physiology addresses the mechanism and response of the body to movement.

Morphology

Morphology, in biology is the study of the size, shape, and structure of animals, plants, and micro-organisms and of the relationships of their constituent parts. The term refers to the general aspects of biological form and arrangement of the parts of a plant or an animal. The term anatomy also refers to the study of biological structure but usually suggests study of the details of either gross or microscopic structure. In practice, however, the two terms are used almost synonymously.

Typically, morphology is contrasted with physiology, which deals with studies of the functions of organisms and their parts; function and structure are so closely interrelated, however, that their separation is somewhat artificial. Morphologists were originally concerned with the bones, muscles, blood vessels, and nerves comprised by the bodies of animals and the roots, stems, leaves, and flower parts comprised by the bodies of higher plants. The development of the light microscope made possible the examination of some structural details of individual tissues and single cells; the development of the electron microscope and of methods for preparing ultrathin sections of tissues created an entirely new aspect of morphology—that involving the detailed structure of cells. Electron microscopy has gradually revealed the amazing complexity of the many structures of the cells of plants and animals. Other physical techniques have permitted biologists to investigate the morphology of complex molecules such as hemoglobin, the gas-carrying protein of blood, and deoxyribonucleic acid (DNA), of which most genes are composed. Thus, morphology encompasses the study of biological structures over a tremendous range of sizes, from the macroscopic to the molecular.

A thorough knowledge of structure (morphology) is of fundamental importance to the physician, to the veterinarian, and to the plant pathologist, all of whom are concerned with the kinds and causes of the structural changes that result from specific diseases.

Evidence that prehistoric humans appreciated the form and structure of their contemporary animals has survived in the form of paintings on the walls of caves in France, Spain, and elsewhere. During the early civilizations of China, Egypt, and the Middle East, as humans learned to domesticate certain animals and to cultivate many fruits and grains, they also acquired knowledge about the structures of various plants and animals.

Aristotle was interested in biological form and structure, and his Historia animalium contains excellent descriptions, clearly recognizable in extant species, of the animals of Greece and Asia Minor. He was also interested in developmental morphology and studied the development of chicks before hatching and the breeding methods of sharks and bees. Galen was among the first to dissect animals and to make careful records of his observations of internal structures. His descriptions of the human body, though they remained the unquestioned authority for more than 1,000 years, contained some remarkable errors, for they were based on dissections of pigs and monkeys rather than of humans.

Although it is difficult to pinpoint the emergence of modern morphology as a science, one of the early landmarks was the publication in 1543 of De humani corporis fabrica by Andreas Vesalius, whose careful dissections of human bodies and accurate drawings of his observations revealed many of the inaccuracies in Galen's earlier descriptions of the human body.

In 1661 an Italian physiologist, Marcello Malpighi, the founder of microscopic anatomy, demonstrated the presence of the small blood vessels called capillaries, which connect arteries and veins.

The existence of capillaries had been postulated 30 years earlier by English physician William Harvey, whose classic experiments on the direction of blood flow in arteries and veins indicated that minute connections must exist between them. Between 1668 and 1680, Dutch microscopist Antonie van Leeuwenhoek used the recently invented microscope to describe red blood cells, human sperm cells, bacteria, protozoans, and various other structures.

Cellular components—the nucleus and nucleolus of plant cells and the chromosomes within the nucleus—and the complex sequence of nuclear events (mitosis) that occur during cell division were described by various scientists throughout the 19th century. Organographie der Pflanzen, the great work of a German botanist, Karl von Goebel, who was associated with morphology in all its aspects, remains a classic in the field. British surgeon John Hunter and French zoologist Georges Cuvier were early 19th-century pioneers in the study of similar structures in different animals—i.e., comparative morphology. Cuvier in particular was among the first to study the structures of both fossils and living organisms and is credited with founding the science of paleontology. A British biologist, Sir Richard Owen, developed two concepts of basic importance in comparative morphology—homology, which refers to intrinsic structural similarity, and analogy, which refers to superficial functional similarity. Although the concepts antedate the Darwinian view of evolution, the anatomical data on which they were based became, largely as a result of the work of German comparative anatomist Carl Gegenbaur, important evidence in favour of evolutionary change, despite Owen's steady unwillingness to accept the view of diversification of life from a common origin.

One of the major thrusts in contemporary morphology has been the elucidation of the molecular basis of cellular structure. Techniques such as electron microscopy have revealed the complex details of cell structure, provided a basis for relating structural details to the particular functions of the cell, and shown that certain cellular components occur in a variety of tissues. Studies of the smallest components of cells have clarified the structural basis not only for the contraction of muscle cells but also for the motility of the tail of the sperm cell and the hairlike projections (cilia and flagella) found on protozoans and other cells. Studies involving the structural details of plant cells, although begun somewhat later than those concerned with animal cells, have revealed fascinating facts about such important structures as the chloroplasts, which contain chlorophyll that functions in photosynthesis. Attention has also been focused on the plant tissues composed of cells that retain their power to divide (meristems), particularly at the tips of stems, and their relationship with the new parts to which they give rise. The structural details of bacteria and blue-green algae, which are similar to each other in many respects but markedly different from both higher plants and animals, have been studied in an attempt to determine their origin.

Morphology continues to be of importance in taxonomy because morphological features characteristic of a particular species are used to identify it. As biologists have begun to devote more attention to ecology, the identification of plant and animal species present in an area and perhaps changing in numbers in response to environmental changes has become increasingly significant.

Fundamental Concepts

Homology and Analogy

Homologous structures develop from similar embryonic substances and thus have similar basic structural and developmental patterns, reflecting common genetic endowments and evolutionary relationships. In marked contrast, analogous structures are superficially similar and serve similar functions but have quite different structural and developmental patterns. The arm of a human, the wing of a bird, and the pectoral fins of a whale are homologous structures in that all have similar patterns of bones, muscles, nerves, and blood vessels and similar embryonic origins; each, however, has a different function. The wings of birds and those of butterflies, in contrast, are analogous structures—i.e., both allow flight but have no developmental processes in common.

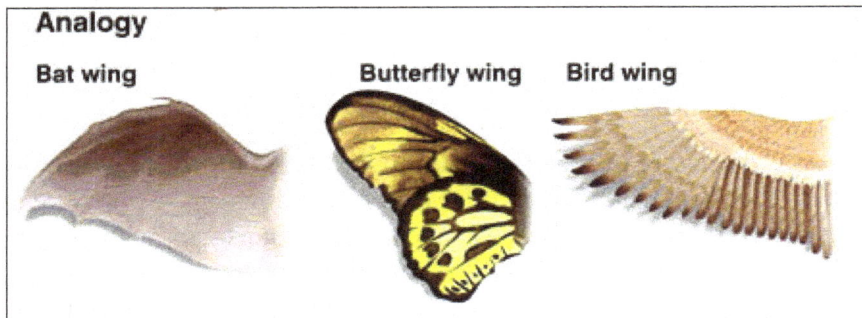

Analogous structure.

Organisms that have evolved along different paths may have analogous structures—that is, anatomical features that are superficially similar to one another (e.g., the wings of birds and insects). Although such structures serve similar functions, they have quite different evolutionary origins and developmental patterns.

The terms homology and analogy are also applied to the molecular structures of cellular constituents. Because the hemoglobin molecules from different vertebrate species contain remarkably similar sequences of amino acids, they may be termed homologous molecules. In contrast, hemoglobin and hemocyanin, the latter of which is present in crab blood, are described as analogous molecules because they have a similar function (oxygen transport) but differ considerably in molecular structure. Corresponding similarities occur in the structures of other proteins from different species—e.g., cytochrome c and other enzymes (biological catalysts) such as the lactic dehydrogenases in birds and mammals.

Body Plan and Symmetry

The bodies of most animals and plants are organized according to one of three types of symmetry: spherical, radial, or bilateral. A spherically symmetrical body is similar throughout and can be cut in any plane through the centre to yield two equal halves. A few of the simplest plants and animals are spherically symmetrical—e.g., protozoans such as Radiolaria and Heliozoa. Radially symmetrical bodies, such as those of starfishes and mushrooms, have a distinguishable top and bottom and usually have a cylindrical shape, with the body parts radiating from the central axis. A starfish can be cut into two equal halves by any plane that includes the line, or axis, running through its centre from top to bottom. The anterior, or oral, end usually contains the mouth; a posterior, or aboral, end may have an anus. In the bilaterally symmetrical body of higher animals including humans, only a cut from head to foot exactly in the centre divides the body into equivalent halves. An anterior, or head, end and a posterior, or tail, end can be distinguished; and the dorsal, or back, side can be distinguished from the ventral, or belly, side. But because some internal organs of humans are not symmetrical (e.g., the heart), even the right and left halves of the human body are not exactly equivalent. A few organisms—amoebas, slime molds, and certain sponges—with an irregular form, or one that changes as the organism moves, have no plane of symmetry.

Morphological basis of Classification

The features that distinguish closely related species of plants and animals are usually superficial differences such as colour, size, and proportion. In contrast, the major divisions, or phyla, of the plant and animal kingdoms are distinguished by characteristics that, though usually not unique to a single division or phylum, occur in unique combinations in each.

One morphological feature useful in classifying animals and in determining their evolutionary relationships is the presence or absence of cellular differentiation—i.e., animals may be either single-celled or composed of many kinds of cells specialized to perform particular functions. Some multicellular animals have only two embryonic cell, or germ, layers: an ectoderm (outer layer) and an endoderm (inner layer), which lines the digestive tract. Other animals have these, in addition to a mesoderm, which lies between the ectoderm and endoderm. Animals may have one of two types of body cavity. The bodies of the Coelenterata (invertebrates such as the jellyfish) and other primitive many-celled animals consist of a double-walled sac surrounding a single cavity with a mouth. Higher animals have two cavities, and their bodies are constructed on a so-called tube-within-a-tube plan. An inner tube, or digestive tract, is lined with endoderm and opens at each end to form the mouth and the anus. An outer tube, or body wall, is covered with ectoderm. Between the two tubes a second cavity, or coelom, lies within the mesoderm and is lined by it. Another major distinguishing morphological feature of animal phyla is the presence or absence of segmentation. The members of several phyla have bodies characterized by the presence of a row of segments, or body units, of the same fundamental structure. Segmented animals include the vertebrates, the annelids (invertebrates such as the earthworm), and the arthropods (invertebrates such as insects); in some segmented animals such as humans and most vertebrates, however, the segmental character of the body is obscured. An evolutionary tendency in many animal phyla has been the progressive differentiation of the anterior end to form a head with conspicuous sense organs and an accumulation of nervous tissues, a brain; the tendency is called cephalization. Some morphological structures are found only in one phylum; for example, only the Coelenterata have stinging cells (nematocysts), the Echinodermata (invertebrates such as starfishes) have a peculiar water vascular system, and only the Chordates (e.g., reptiles, birds) have a dorsally located, hollow nerve cord.

Like animals, plants may be either single-celled or composed of many kinds of specialized cells. The bodies of most of the lower plants, such as algae and fungi, comprise the least-differentiated and least-specialized type of plant cells, parenchyma cells. The embryonic tissues of higher plants, unlike those of animals, remain extremely active throughout the life of the plant. In addition, the different types of cells characteristic of the body of higher plants arise from meristems, specific regions in the plant body where cells divide and enlarge. In all but the simplest forms, the plant body is composed of various types of cells associated in more or less definite ways to form systems of units called tissue systems—e.g., the vascular system consisting of conductive tissues. The arrangement of the components of the vascular system is a distinguishing morphological feature of various plant groups. The character and relative extent of the two phases in the life history of a plant—the sexual phase, or gametophyte, and the sporophyte—vary considerably among the plant groups and are useful in distinguishing them.

Areas of Study

Anatomy

The best known aspect of morphology, usually called anatomy, is the study of gross structure, or form, of organs and organisms. It should not be inferred however, that even the human body, which has been extensively studied, has been so completely explored that nothing remains to be discovered. It was found only in 1965, for example, that the nerve to the pineal gland, which lies on the upper surface of the brain of mammals, is a branch from the sympathetic nerves; the sympathetic nerves receive nerve impulses from a small branch of the nerves that transmit impulses from the eye to the brain (optic nerves). Thus the pineal gland responds by a very indirect route to quantitative changes in the environmental lighting and secretes appropriate amounts of the substance it forms, the hormone melatonin.

Detailed comparisons of the morphological features of different animals, called comparative anatomy, provide strong arguments for the evolutionary relationships among different species. In the course of evolution, animals and plants tend to undergo adaptive morphological changes that enable them to survive under certain environmental conditions. As a result, animals only remotely related evolutionarily may come to resemble each other superficially because of common adaptations to similar environments, a phenomenon known as convergent evolution. Structural similarities—streamlined shape, dorsal fins, tail fins, and flipper-like forelimbs and hindlimbs, for example—have evolved in such varied animal groups as the dolphins and porpoises, both of which are mammals; the extinct ichthyosaurs, which were reptiles; and both the bony and cartilaginous fishes. In a like manner, the mole, an insectivore, and the gopher, a rodent, have both evolved shovellike forelimbs, an adaptation for digging.

An opposite phenomenon, divergent evolution, occurs when animals originally closely related adapt to different environments and come to be superficially quite different. Although sea lions and seals, for example, are carnivores and thus closely related to bears, cats, and dogs, their adaptations to an aquatic existence have resulted in morphological characteristics distinct from those of the terrestrial carnivores. In the course of mammalian evolution, many features have changed to permit specific animal groups to adapt to particular environments—e.g., the number and shape of the teeth, the length and number of bones in the limbs, the number and attachment sites of muscles, the thickness and colour of the hair or fur, and the length and shape of the tail.

Careful study of adaptive morphological aspects has permitted inferences about the course of the evolutionary history of various animals and of their successive adaptations to changing environments. The present-day Australian tree-climbing kangaroos, for example, are the descendents of a ground-dwelling marsupial, from whom evolved forms that began to live in trees and eventually developed limbs adapted to tree climbing. But the events may have occurred in the reverse sequence; that is, specialized limbs may have evolved before the animal adopted an arboreal mode of life. In any event, some of the tree-dwelling kangaroos subsequently left the trees, became readapted to life on the ground (i.e., their hindlegs became adapted for leaping), and then went back to the trees but with legs so highly specialized for leaping as to be useless in grasping a tree trunk; consequently, present-day tree kangaroos climb by bracing their feet against a tree trunk, as do bears. Careful comparisons of the feet of the many kinds of living Australian marsupials reveal the stages in this complicated process of adaptation and re-adaptation.

Changes in genes (mutations) constantly occur and may cause a decrease in size and function of an organ. On the other hand, a change in the environment or in the mode of life of a species may make an organ unnecessary for survival. As a result, many plants and animals contain organs or parts of organs that are useless, degenerate, undersized, or lacking some essential part when compared with homologous structures in related organisms. The human body, for instance, has more than 100 such organs—e.g., the appendix, the fused tail vertebrae (coccyx), the wisdom teeth, the muscles that wiggle the ears, and the hair on the body.

The parts of a seed plant include roots, stems, leaves, and reproductive organs in the flowers. The evolution of specialized conducting tissues called xylem and phloem has enabled seed plants to survive on land and to attain large sizes. Roots anchor the plant, enable it to maintain an upright position, and absorb water, minerals, and other nutrients from the soil. The roots of plants such as carrots, beets, and yams serve as sites for food storage. The stem links the roots with the leaves, where photosynthesis occurs, and its xylem and phloem are continuous with those of root and leaf. The stem supports leaves, flowers, and fruits. Each year, the stems of woody plants add a layer of xylem and phloem, the annual ring, the width of which varies with climatic conditions. A leaf consists of a petiole (stalk), by which it is attached to the stem, and a blade, typically broad and flat, that contains bundles, or veins, of xylem and phloem on the undersurface. The flower contains pollen-producing anthers and egg-producing ovules. After fertilization the base of the flower, or ovary, enlarges and forms the fruit, which is a mature ovary containing seeds, or mature ovules. The bodies of ferns and mosses also are composed of roots, stems, and leaves, but those of lower plants such as mushrooms and kelps are much more simple and lack true roots, stems, and leaves.

A major trend in the evolution of both plants and animals has resulted in the specialization of cells and a division of labour among them. The cells that make up a tree or a human are quite different; each is specialized to carry out certain functions. Although specialization may permit a cell to function efficiently, it also increases the interdependence of body parts; an injury to or the destruction of one part, therefore, may result in death of the whole organism. The study of the structure and arrangement of tissues, defined as groups or layers of cells that together perform certain special functions, is known as histology. Each kind of tissue is composed of cells with characteristic features such as size, shape, and relationship to adjacent cells and may also contain noncellular material—connective tissue fibres or a bony material.

Morphologists usually separate animal tissues into six groups: epithelial, connective, muscular, blood, nervous, and reproductive tissues. The cells of epithelial tissues form a continuous layer or sheet that either covers the surface of the body or lines some cavity within the body, thus protecting the underlying cells from mechanical and chemical injury or from invasion by microorganisms. Epithelial tissues absorb nutrients and water, secrete a wide variety of substances, and may play a role in the reception of sensory stimuli. The connective tissues—bone, cartilage, ligaments, and fibrous connective tissue—support and hold together the other cells of the body. The cells of the connective tissues secrete large quantities of nonliving material (matrix), the characteristics of which largely determine the nature and the function of the specific types of connective tissue; the matrix secreted by fibrous connective tissue cells, for example, is a thick matted network of microscopic fibres surrounding the connective tissue cells. Connective tissue holds skin to muscle, keeps glands in position, makes up the tough outer walls of the blood vessels, and forms a sheath around nerve fibres and muscle cells. Tendons are flexible, cable-like cords of specialized fibrous connective tissue that join muscles to each other or muscle to bone. Ligaments are somewhat elastic cords of specialized fibrous connective tissue that join one bone to another.

Muscular tissues are composed of elongated, cylindrical, or spindle-shaped cells, each of which contains many small fibres called myofibrils. Muscle cells perform mechanical work by contracting—that is, by becoming shorter and thicker. The three types of vertebrate muscles include the cardiac muscle, which is found only in the walls of the heart; smooth muscles, which are found in the walls of the digestive tract and in other internal organs; and skeletal muscles, which make up the bulk of the muscle masses attached to the bones of the body. Skeletal and cardiac muscles have alternating light and dark stripes the relative sizes of which change during the contraction process. Evidence from electron microscopy indicates that two types of filaments occur in muscle; during contraction, one type of filament slides past the other.

Nerve tissue is made of cells, called neurons, which are specialized to conduct nerve impulses. Two or more thin hairlike fibres, called axons and dendrites, extend from the enlarged cell body containing the nucleus. The neurons extending from the spinal cord to the end of an appendage (e.g., arm, leg) may extend to a metre (about three feet) or more in humans and to several metres in an elephant or a whale.

Egg cells in the female and sperm cells in the male are reproductive tissues adapted for the production of offspring. The egg cell is modified by the accumulation of considerable amounts of yolk and other food reserves. The highly specialized spermatozoon contains a tail, the beating of which propels it to the egg.

Blood is composed of red cells, which are specialized for the transport of oxygen and carbon dioxide, and white cells, which engulf bacteria and produce antibodies (proteins formed in response to foreign substances called antigens). Blood also contains platelets, small fragments of cells from the bone marrow that play a key role in initiating the clotting of blood.

The cells of higher plants may be differentiated into meristematic, protective, fundamental, and conductive tissues. Meristematic tissues, which are composed of small thin-walled cells with few or no vacuoles (cavities), differentiate into the other types of plant tissue and are found in the rapidly growing parts of the plant—e.g., at the tips of roots and stems. Protective tissues are composed of thick-walled cells that protect the underlying thin-walled cells from mechanical abrasion and dehydration; examples of protective tissues include the epidermis of leaves and the cork layers of stems and roots. The fundamental tissues that constitute the body of a plant include the soft parts of the leaf, the components of the pith and the cortex of stems, the roots, and the soft parts of flowers and fruits. These tissues function in the production and storage of food. Two types of conductive tissues occur in higher plants: xylem conducts water and dissolved salts, and phloem conducts dissolved organic materials such as sugars. Both types are composed of elongated cells that fuse end to end with other cells to form the sieve tubes through which substances are transported in phloem and xylem vessels.

Cytology

The living material of most organisms is organized into discrete units called cells, and the study of their features is known as cytology. The cellular contents, when viewed through a microscope at low magnification, usually appear to consist of granules or fibrils of dense material, droplets of fatty substances, and fluid-filled vacuoles suspended in a clear, continuous, semifluid substance called cytoplasm. The remarkable structural complexity of the cell is more fully revealed at the higher magnifications attainable with the electron microscope. Structural details of various cellular components, or organelles, as revealed by the technique known as X-ray diffraction analysis, have provided information concerning the relationships between the structures of the cellular components and of the molecules that constitute them. Although most cells have certain features in common, the kinds and amounts of components vary considerably. Cellular components include structures such as mitochondria, chloroplasts, endoplasmic reticulum, Golgi complex, lysosomes, oil droplets, granules, and fibrils. The cell is surrounded by a membrane, and similar membranes surround many cellular components—e.g., the mitochondria.

A small spherical or oval organelle, the nucleus, is typically found near the centre of a cell. The genes within the nucleus control the development of the various traits of the cell by controlling the synthesis of specific proteins. The nuclear components are separated from those of the cytoplasm by the nuclear membrane. The structure of the nucleolus, a spherical body within the nucleus, is extremely variable in most cells. Although more than one nucleolus may occur in a nucleus, each cell of an animal or plant species has a fixed number of nucleoli. The nucleoli apparently play a role in the synthesis of the ribonucleic acid (RNA) constituent of the cellular components called ribosomes, which function in protein synthesis. Adjacent to the nucleus in the cells of animals and certain lower plants are two small, cylindrical bodies, the centrioles, which, during cell division, separate, migrate to opposite sides of the cell, and organize a structure called a spindle between them.

Within the cytoplasm of both plant and animal cells are components called mitochondria, which may be shaped like spheres, rods, or threads. Each mitochondrion is bounded by a double membrane, the outer layer of which forms the smooth outer boundary of the mitochondrion; the inner layer, folded repeatedly into shelflike folds called cristae, contains enzymes that play an essential role in the conversion of the energy of foodstuffs into the energy used for cellular activities. The cells of most plants contain plastids, small bodies involved in the synthesis and storage of foodstuffs. The most important plastids, the chloroplasts, function in trapping the energy of sunlight during photosynthesis. They are disk-shaped structures with a platelike arrangement of tightly stacked membranes.

The cytoplasmic components important in protein synthesis, the ribosomes, are composed of nucleic acid and protein. Clusters of five or more ribosomes, termed polysomes, appear to be the functional unit in protein synthesis.

Lysosomes are membrane-bound structures containing a variety of enzymes that can break down the large molecular constituents of the cell. The membrane surrounding lysosomes presumably prevents the enzymes from digesting the cell contents before the cell dies.

Embryology

The structures and the relationships among the various parts of a mature plant or animal are usually better understood if the successive developmental stages are studied. Thus, morphologists have traditionally been interested in the study of embryos and their developmental patterns—i.e., the science of embryology.

Development typically begins in animals with the cleavage, or division, of the fertilized egg (zygote) to form a hollow ball of cells called the blastula; the blastula then develops into a hollow cuplike body of two layers of cells, the gastrula, from which the embryo ultimately is formed. At one time, the techniques available to embryologists enabled them to study only whole embryos at different developmental stages. The science of experimental embryology began during the first half of the 20th century, when microsurgical techniques became available either for the removal and study of certain structures from tiny embryos or for their transplantation to other regions of the embryo. Advances in understanding the mechanism by which biological information is transferred in DNA and the means by which this information results in the production of specific proteins have led to efforts to describe development in biochemical terms. Although hypotheses regarding the reasons for the appearance of a specific enzyme or some other protein at a specific time during development have been formulated and tested, the biochemical basis of morphogenesis itself—that is, the reason for the development of particular structures—is not fully understood.

The development of the seed plant is basically different from that of an animal. The egg cell of a seed plant is retained within the enlarged lower part, or ovary, of the seed-bearing organ (pistil) of a flower. Two sperm nuclei pass through a structure called a pollen tube to reach the egg. One sperm nucleus unites with the egg nucleus to form the zygote from which the new plant will develop. The second sperm nucleus unites with two nuclei, called polar nuclei, to form a body called a triploid endosperm, the cells of which divide to form a nutritive mass within the seed. The zygote undergoes several cell divisions to form the embryo, which is surrounded by the endosperm. The embryo develops one or two seed leaves, or cotyledons, which may become thick and fleshy with stored foodstuffs. The epicotyl, which consists of a growing point enclosed by a pair of folded miniature leaves, develops above the point of attachment of the seed leaves. Below the seed leaves extends the hypocotyl, the tip, or radicle, of which forms the primary root of the embryonic plant.

The factors involved in initiating and controlling morphogenesis in plants have been studied by growing cells, tissues, and organs derived from plants. Indeed, an entire carrot plant has been grown from one cell of a mature carrot. This provides striking evidence that the cell from the adult plant contains all of the genetic information needed to produce an entire plant, including roots, stems, and leaves. The technique of growing plants from isolated plant parts has been useful in studies involving the characteristics of embryonic growth, the correlated growth of plant parts, and the nature of differentiation and regeneration (the replacement of lost parts).

Methods In Morphology

Chemical Techniques

The methods of investigating gross structure depend on careful dissection, or cutting apart, of an organism and on accurate descriptions of the parts. The study of the structure of tissues and cells has been extended by the techniques of autoradiography and histochemistry. In the former, a tissue is supplied with a radioactive substance and allowed to utilize it for an appropriate period of time, after which the tissue is prepared and placed in contact with a special photographic emulsion. Silver grains in the emulsion in contact with radioactive substances darken; thus, the location of the dark spots indicates the position at which the radioactive substance was concentrated in the tissue. Histochemistry involves the differential staining of cells (i.e., using dyes that stain specific structural and molecular components) to reflect the chemical differences of the constituents. By choosing appropriate dyes, the histochemist is able, for example, to determine the acidity or alkalinity of the chemical compounds that make up cell components. In addition, dyes that stain specific molecular constituents such as glycogen, DNA, RNA, and protein also are used. The histochemist is able to locate a specific enzyme in a thin slice of tissue, to provide the specific substance with which the enzyme reacts to form a product, and to add a compound that reacts with the product to form an insoluble coloured compound the location of which is relatively easy to determine In this way, information has been obtained about the specific location of enzymes within the cell.

Microscopic Techniques

Histologists and cytologists utilize microscopic techniques—light microscopy, phase contrast microscopy, interference microscopy, polarization microscopy, fluorescent microscopy, and electron microscopy—to investigate certain aspects of cell structure. Phase contrast microscopy is widely used to study the structure of living cells because, with such apparatus, internal structures can be observed without killing and staining the cell. In addition, motion pictures of dividing cells or moving cells can be made using phase contrast microscopy.

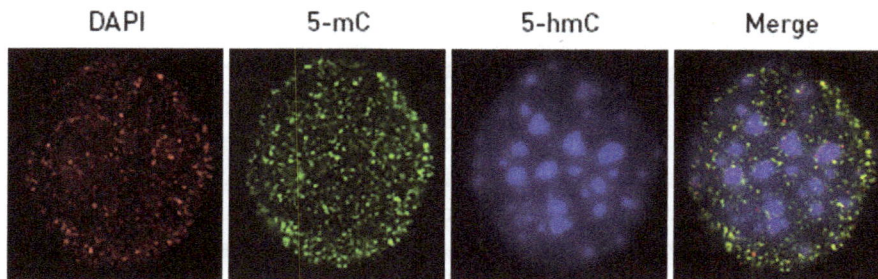

Immunoflourescence Immunofluorescence results obtained with 5-hmC monoclonal antibody. NIH 3T3 cells (mouse embryonic fibroblasts) have been grown on coverslips.

The interference microscope involves passing two separate beams of light through the specimen. With the appropriate instrument, the mass of material per unit area of the specimen can be determined, and contour mapping of small objects is possible.

Crystalline or fibrous elements, both of which are characterized by an orderly or layered molecular structure, are studied with a polarizing microscope; the polarizing microscope has been particularly useful in studying the detailed structure of bone.

In fluorescence microscopy, the images seen are molecules of fluorescent dyes added to cells that attach to specific cellular components. Appropriate filters are required to insure that only the light of longer wavelength contributes to the image. Fluorescent antibodies have been used to locate specific kinds of proteins and other materials in certain cells of a tissue or in certain regions of a cell. The antibodies are prepared by injecting into a rabbit an antigen (e.g., the protein myosin), which stimulates white blood cells called lymphocytes to synthesize antibodies that react specifically with the antigen. After the antibodies are isolated and purified, the fluorescent dye, fluorescein, becomes attached to them by a chemical reaction. If the fluorescent antibodies are spread over a tissue, they attach specifically to the molecules that stimulated their formation (myosin).

The fluorescence microscope reveals the sites containing the antigen–antibody complex as bright luminescent areas in a dark background.

In the scanning electron microscope, a moving spot of electrons (negatively charged particles) is used to scan an object and to produce an image similar to that which appears on a television screen. In this manner, photographs with a three-dimensional appearance can be produced. With the transmission electron microscope, a beam of electrons passes through an object, such as a cell, and is focused on the other side onto a fluorescent screen or a photographic plate. The beam of electrons in the scanning electron microscope is focused and then scanned across the specimen. The electrons that leave the specimen, which are not necessarily the same electrons that strike it, are then used to control the beam of a cathode-ray picture tube. Scanning electron microscopes allow photographs to be taken not only of large molecules such as DNA but of very small objects— individual atoms of elements such as uranium or thorium.

2

Vertebrate Physiology

The branch of biology which studies the functions and mechanisms inside vertebrates is termed as vertebrate physiology. Some of the important types of vertebrates studied within this field are birds and fish. This chapter closely examines the key concepts related to the physiology of these vertebrates to provide an extensive understanding of the subject.

Vertebrates are animals belonging to the subphylum Vertebrata, that is, animals with backbones or spinal columns. Additional defining characteristics of the subphylum are a muscular system that mostly consists of paired masses, as well as a central nervous system that is partly located inside the backbone. The name of this group comes from the bones of the spinal column (or vertebral column), which are called vertebrae.

Vertebrata is the largest subphylum of the phylum Chordata (chordates), and includes animals with which many people are familiar. Fish (including lampreys), amphibians, reptiles, birds, and mammals (including humans) are vertebrates. Over 50,000 species of vertebrates have been described. However, more than 95 percent of described animal species are invertebrates—a disparate classification of all animals without backbones.

In general, people feel a strong affinity toward vertebrates. Although invertebrates greatly outnumber vertebrates, it is overwhelmingly vertebrates that are kept as pets or in aquariums and terrariums, as well as being featured in movies, animated films, and other media. Communities pass laws relating to the humane treatment of vertebrates, and they build zoos for their display. The more similar the class of vertebrates is to humans, generally the more the apparent attraction and the more laws for their protection—as seen in the special attention shown to birds and especially mammals, versus fish, amphibians, and reptiles. (For example, some universities do not permit research on any vertebrates other than fish, amphibians, and reptiles.) This may be a reflection of the philosophical perspective that people can best love and feel joy from those beings that can most mirror their own character and form.

The study of animals with backbones is called vertebrate zoology. Vertebrate paleontology is the study of animals with backbones through their fossilized remains, and includes studies aimed at connecting animals of the past with modern day relatives.

Characteristic of Vertebrates

In addition to the spinal cord, central nervous system, and unique muscular system, vertebrates also are characterized by an internal skeleton and a brain case.

The internal skeleton that defines vertebrates consists of cartilage or bone, or in some cases both. It is speculated that the first bony substance that vertebrates evolved was an outer skeleton in the form of a bony armor, and that its primary function was as a phosphate reservoir, excreted as calcium phosphate and stored around the body, offering protection at the same time. The internal skeleton provides support to the organism during the period of growth. For this reason, vertebrates can achieve larger sizes than invertebrates, and on average vertebrates are, in fact, larger. The skeleton of most vertebrates, excluding the most primitive ones, consists of a skull, the vertebral column, and two pairs of limbs. In some forms of vertebrates, one or both of these pairs of limbs may be absent, such as in snakes or whales. These limbs are assumed to have been lost during the course of evolution.

Northern Bluefin Tuna, Thunnus thynnus.

The skull is thought to have facilitated the development of intelligence as it protects vital organs such as the brain, the eyes, and the ears. The protection of these organs is also thought to have positively influenced the development of the high responsiveness to the environment often found in vertebrates.

Both the vertebral column and the limbs offer overall support to the body of the vertebrate. This support facilitates movement, which is normally achieved with muscles that are attached directly to the bones or cartilages. The muscles form the contour of the body of a vertebrate. A skin covers the inner parts of a vertebrate's body. The skin sometimes acts as a structure for protective features, such as horny scales or fur. Feathers may also be attached to the skin.

The trunk of a vertebrate houses the internal organs. The heart and the respiratory organs are protected in the trunk. The heart is located either behind the gills, or, in air-breathing vertebrates, between the lungs.

The central nervous system of a vertebrate consists of the brain and the spinal cord. In lower vertebrates, the brain mostly controls the functioning of the sense organs. In higher vertebrates, the size of the brain relative to the size of the body generally is greater. This larger brain enables more intensive exchange of information between the different parts of the body. The nerves from the spinal cord, which lies behind the brain, extend to the skin, the inner organs, and the muscles. Some nerves are directly connected to the brain, linking the brain with the ears and lungs.

A 2012 article in the scholarly publication Science identified the world's smallest known vertebrate as the frog species Paedophryne amauensis, with adults reaching a average size of 7.7 millimeters in length. Its discovery in a rainforest in eastern New Guinea pushed into second place an Indonesian fish from the carp family, with the adult female fish growing to about 7.9 millimeters.

The world's largest vertebrate is the blue whale (Balaenoptera musculus), which reaches up to 33 meter (110 ft) in length and 181 metric ton (200 short tons) or more in weight.

Number of Species of Vertebrates

Vertebrates are the best known among the animals, with most species having been identified and described. There are comprehensive listings of the extant (living) species of mammals, birds, reptiles, amphibians, and fish.

Nonetheless, the determination of the number of species of vertebrates is necessarily imprecise. One reason is that taxonomists generally strive to arrange species based on evolutionary relationships. As more insights are obtained, there are taxonomic rearrangements and new nomenclature, even to the point that sometimes species are reclassified as subspecies, and vice-versa. For example, in herpetological classifications, dealing with reptiles and amphibians, the adoption of the evolutionary species concept, versus the previously used biological species concept, led to the elevation of many subspecies to species status. Molecular studies are expected to lead to additional rearrangements. A second reason why it is difficult to determine the exact number of species is that new species are continually being discovered and described. Fish are being described at a rate of about 200 per year, amphibians at the rate of about 80 species per year, and reptiles at the rate of about 60 species per year.

According to a report by Uetz in 2000, comprehensive compilations of vertebrates reveal a species total of 4,675 mammals, 9,702 birds, 7,870 reptiles, 4,780 amphibians, and 23,250 fishes. Of the reptiles, the majority were determined to be lizards (4,470 species) and snakes (2,920). Over one half of all reptile species fall into the category of either colubrid snakes (approximately 1,800 species), skinks (1,200 species), or geckos (1,000 species). A subsequent tabulation by Uetz in 2005 showed a total of 8,240 extant reptile species.

A 2004 list of species by the International Union for the Conservation of Nature and Natural Resources (IUCN) yielded the following number of described species of vertebrates: mammals (5,416), birds (9,917), reptiles (8,163), amphibians (5,743), and fishes (28,500). This totals to 57,739 identified vertebrate species. Meanwhile, the IUCN reports a total of 1,190,200 described, extant species of invertebrates (although this represents an assumably small proportion of actual species of this very incompletely known group).

Vertebrates have been traced back to the ostracoderms (primitive jawless fish) of the Silurian Period (444 million to 409 million years ago) and the conodonts, a group of eel-like vertebrates characterized by multiple pairs of bony toothplates. Vertebrates started to evolve about 530 million years ago during the Cambrian explosion.

Taxonomy

Vertebrates (subphylum Vertebrata) are part of the phylum Chordata, which are animals that had, at some stage in their life, a notochord, a hollow dorsal nerve cord, and pharyngeal slits, among other characteristics. Chordata includes two subphyla of invertebrates (Urochordata and Cephalochordata) and the Vertebrates.

Vertebrates are also considered part of the Craniata, a group of animals that includes all animals with skulls. Craniata consists of the vertebrates and hagfish (Myxini). In some taxonomies, hagfish, which lack vertebrae, nonetheless are included in Vertebrata based on presumed evolutionary relatedness.

Bird Anatomy

Bird anatomy, or the physiological structure of birds' bodies, shows many unique adaptations, mostly aiding flight. Birds have a light skeletal system and light but powerful musculature which, along with circulatory and respiratory systems capable of very high metabolic rates and oxygen supply, permit the bird to fly. The development of a beak has led to evolution of a specially adapted digestive system. These anatomical specializations have earned birds their own class in the vertebrate phylum.

Skeletal System

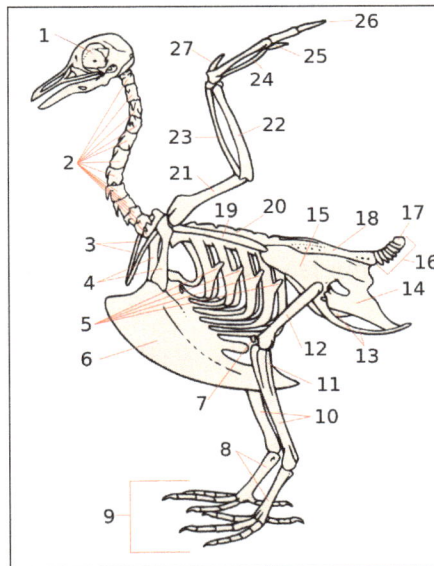

A stylised dove skeleton: 1. skull 2. cervical vertebrae 3. furcula 4. coracoid 5. uncinate processes of ribs 6. keel 7. patella 8. tarsometatarsus 9. digits 10. tibia (tibiotarsus) 11. fibula (tibiotarsus) 12. femur 13. ischium (innominate) 14. pubis (innominate) 15. illium (innominate) 16. caudal vertebrae 17. pygostyle 18. synsacrum 19. scapula 20. dorsal vertebrae 21. humerus 22. ulna 23. radius 24. carpus (carpometacarpus) 25. metacarpus (carpometacarpus) 26. digits 27. alula.

Birds have many bones that are hollow (pneumatized) with criss-crossing struts or trusses for structural strength. The number of hollow bones varies among species, though large gliding and soaring birds tend to have the most. Respiratory air sacs often form air pockets within the semi-hollow bones of the bird's skeleton. The bones of diving birds are often less hollow than those of non-diving species. Penguins, loons, and puffins are without pneumatized bones entirely. Flightless birds, such as ostriches and emus, have pneumatized femurs and, in the case of the emu, pneumatized cervical vertebrae.

Axial Skeleton

The bird skeleton is highly adapted for flight. It is extremely lightweight but strong enough to withstand the stresses of taking off, flying, and landing. One key adaptation is the fusing of bones into single ossifications, such as the pygostyle. Because of this, birds usually have a smaller number of bones than other terrestrial vertebrates. Birds also lack teeth or even a true jaw, and instead have a beak, which is far more lightweight. The beaks of many baby birds have a projection called an egg tooth, which facilitates their exit from the amniotic egg, which falls off once it has done its job.

Vertebral Column

The vertebral column is divided into five sections of vertebrae:

- Cervical (11–25) (neck).

- Trunk (dorsal or thoracic) vertebrae usually fused in the notarium.

- Synsacrum (fused vertebrae of the back also fused to the hips/pelvis). This region is similar to the sacrum in mammals and is unique in the pigeon because it is a fusion of the sacral, lumbar, and caudal vertebra. It is attached to the pelvis and supports terrestrial locomotion of the pigeon's legs.

- Caudal (5–10): This region is similar to the coccyx in mammals and helps control the movement of feathers during flight.

- Pygostyle (tail): This region is made up of 4 to 7 fused vertebrae and is the point of feather attachment.

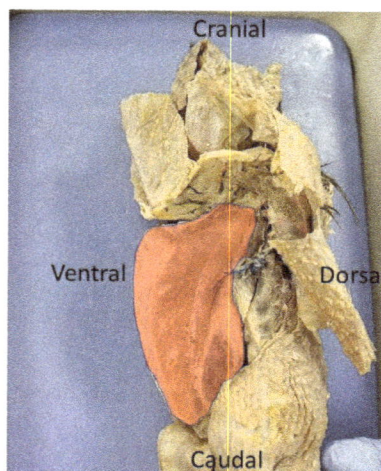

Highlighted in red is an intact keeled sternum of a dissected pigeon. In flying birds the sternum is enlarged for increased muscle attachment.

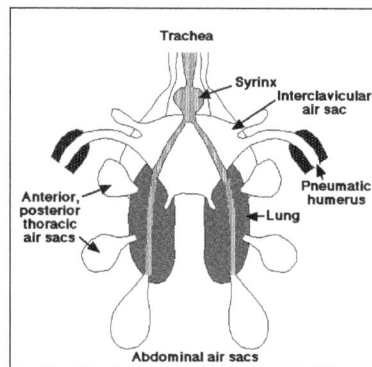

Air-sacs and their distribution.

The neck of a bird is composed of 13–25 cervical vertebrae enabling birds to have increased flexibility. A flexible neck allows many birds with immobile eyes to move their head more productively and center their sight on objects that are close or far in distance. Most birds have about three times as many neck vertebrae as humans, which allows for increased stability during fast movements such as flying, landing, and taking-off. The neck plays a role in head-bobbing which is present in at least 8 out of 27 orders of birds, including Columbiformes, Galliformes, and Gruiformes. Head-bobbing is an optokinetic response which stabilizes a birds surroundings as they alternate between a thrust phase and a hold phase. Head-bobbing is synchronous with the feet as the head moves in accordance with the rest of the body. Data from various studies suggest that the main reason for head-bobbing in some birds is for the stabilization of their surroundings, although it is uncertain why some but not all bird orders show head-bob.

Birds are the only vertebrates to have fused collarbones and a keeled breastbone. The keeled sternum serves as an attachment site for the muscles used in flying or swimming. Flightless birds, such as ostriches, lack a keeled sternum and have denser and heavier bones compared to birds that fly. Swimming birds have a wide sternum, walking birds have a long sternum, and flying birds have a sternum that is nearly equal in width and height.

The chest consists of the furcula (wishbone) and coracoid (collar bone), which, together with the scapula, form the pectoral girdle. The side of the chest is formed by the ribs, which meet at the sternum (mid-line of the chest).

Ribs

Birds have uncinate processes on the ribs. These are hooked extensions of bone which help to strengthen the rib cage by overlapping with the rib behind them. This feature is also found in the tuatara (*Sphenodon*).

Skull

The skull consists of five major bones: the frontal (top of head), parietal (back of head), premaxillary and nasal (top beak), and the mandible (bottom beak). The skull of a normal bird usually weighs about 1% of the bird's total body weight. The eye occupies a considerable amount of the skull and is surrounded by a sclerotic eye-ring, a ring of tiny bones. This characteristic is also seen in reptiles.

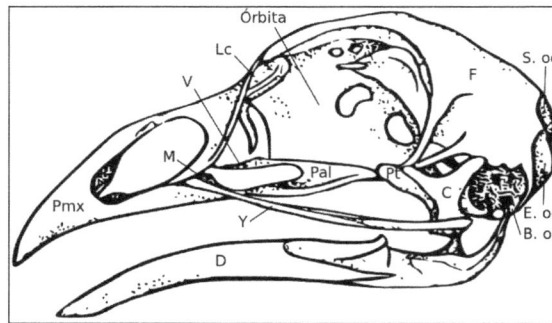

The typical cranial anatomy of a bird. Pmx= premaxilla, M= maxilla,
D= dentary, V= vomer, Pal= palatine, Pt= Pterygoid, Lc= Lacrimal.

Broadly speaking, avian skulls consist of many small, non-overlapping bones. Paedomorphosis, maintenance of the ancestral state in adults, is thought to have facilitated the evolution of the avian skull. In essence, adult bird skulls will resemble the juvenile form of their theropod dinosaur ancestors. As the avian lineage has progressed and has paedomorphosis has occurred, they have lost the postorbital bone behind the eye, the ectopterygoid at the back of the palate, and teeth. The palate structures have also become greatly altered with changes, mostly reductions, seen in the ptyergoid, palatine, and jugal bones. A reduction in the adductor chambers has also occurred These are all conditions seen in the juvenile form of their ancestors. The premaxillary bone has also hypertrophied to form the beak while the maxilla has become diminished, as suggested by both developmental and paleontological studies. This expansion into the beak has occurred in tandem with the loss of a functional hand and the developmental of a point at the front of the beak that resembles a "finger". The premaxilla is also known to play a large role in feeding behaviors in fish.

The structure of the avian skull has important implications for their feeding behaviors. Birds show independent movement of the skull bones known as cranial kinesis. Cranial kinesis in birds occurs in several forms, but all of the different varieties are all made possible by the anatomy of the skull. Animals with large, overlapping bones (including the ancestors of modern birds) have akinetic (non-kinetic) skulls. For this reason it has been argued that the paedomorphic bird beak can be seen as an evolutionary innovation.

Birds have a diapsid skull, as in reptiles, with a pre-lachrymal fossa (present in some reptiles). The skull has a single occipital condyle.

Appendicular Skeleton

The shoulder consists of the scapula (shoulder blade), coracoid, and humerus (upper arm). The humerus joins the radius and ulna (forearm) to form the elbow. The carpus and metacarpus form the "wrist" and "hand" of the bird, and the digits are fused together. The bones in the wing are extremely light so that the bird can fly more easily.

The hips consist of the pelvis, which includes three major bones: the ilium (top of the hip), ischium (sides of hip), and pubis (front of the hip). These are fused into one (the innominate bone). Innominate bones are evolutionary significant in that they allow birds to lay eggs. They meet at the acetabulum (hip socket) and articulate with the femur, which is the first bone of the hind limb.

The upper leg consists of the femur. At the knee joint, the femur connects to the tibiotarsus (shin) and fibula (side of lower leg). The tarsometatarsus forms the upper part of the foot, digits make up the toes. The leg bones of birds are the heaviest, contributing to a low center of gravity, which aids in flight. A bird's skeleton accounts for only about 5% of its total body weight.

They have a greatly elongate tetradiate pelvis, similar to some reptiles. The hind limb has an intra-tarsal joint found also in some reptiles. There is extensive fusion of the trunk vertebrae as well as fusion with the pectoral girdle.

Feet

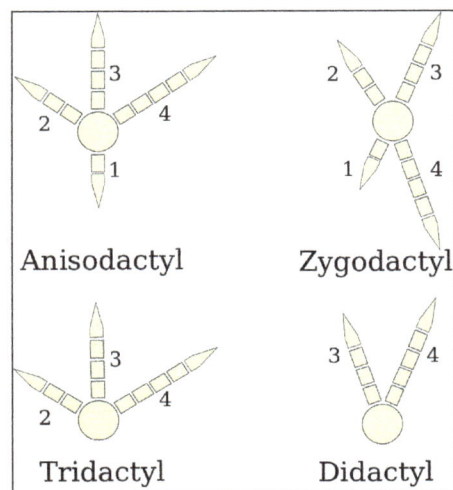

Four types of bird feet (right foot diagrams).

Birds' feet are classified as anisodactyl, zygodactyl, heterodactyl, syndactyl or pamprodactyl. Anisodactyl is the most common arrangement of digits in birds, with three toes forward and one back. This is common in songbirds and other perching birds, as well as hunting birds like eagles, hawks, and falcons.

Syndactyly, as it occurs in birds, is like anisodactyly, except that the third and fourth toes (the outer and middle forward-pointing toes), or three toes, are fused together, as in the belted kingfisher *Ceryle alcyon*. This is characteristic of Coraciiformes (kingfishers, bee-eaters, rollers, etc.).

Zygodactyl feet have two toes facing forward (digits two and three) and two back (digits one and four). This arrangement is most common in arboreal species, particularly those that climb tree trunks or clamber through foliage. Zygodactyly occurs in the parrots, woodpeckers (including flickers), cuckoos (including roadrunners), and some owls. Zygodactyl tracks have been found dating to 120–110 Ma (early Cretaceous), 50 million years before the first identified zygodactyl fossils.

Heterodactyly is like zygodactyly, except that digits three and four point forward and digits one and two point back. This is found only in trogons, while pamprodactyl is an arrangement in which all four toes may point forward, or birds may rotate the outer two toes backward. It is a characteristic of swifts (Apodidae).

Muscular System

Most birds have approximately 175 different muscles, mainly controlling the wings, skin, and legs. Overall, the muscle mass of birds is concentrated ventrally. The largest muscles in the bird are the pectorals, or the pectoralis major, which control the wings and make up about 15–25% of a flighted bird's body weight.They provide the powerful wing stroke essential for flight. The muscle deep to (underneath) the pectorals is the supracoracoideus, or the pectoralis minor. It raises the wing between wingbeats. Both muscle groups attach to the keel of the sternum. This is remarkable, because other vertebrates have the muscles to raise the upper limbs generally attached to areas on the back of the spine. The supracoracoideus and the pectorals together make up about 25–40% of the bird's full body weight. Caudal to the pectorals and supracoracoides are the internal and external obliques which compress the abdomen. Additionally, there are other abdominal muscles present that expand and contract the chest, and hold the ribcage. The muscles of the wing, as seen in the labelled images, function mainly in extending or flexing the elbow, moving the wing as a whole or in extending or flexing particular digits. These muscles work to adjust the wings for flight and all other actions. Muscle composition does vary between species and even within families.

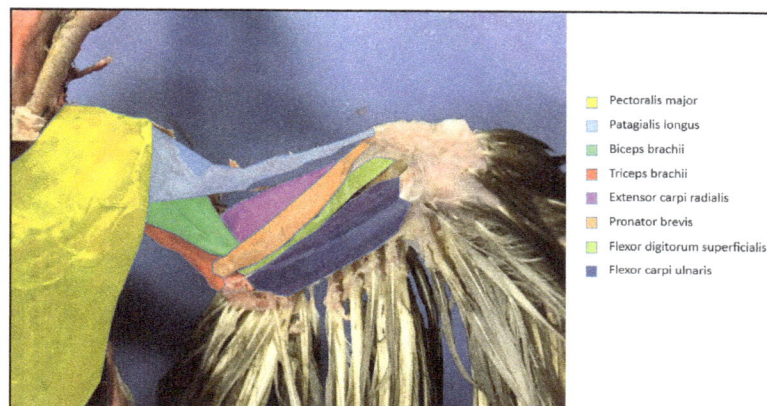

Labelled ventral musculature of a pigeon wing.

Birds have unique necks which are elongated with complex musculature as it must allow for the head to perform functions other animals may utilize pectoral limbs for.

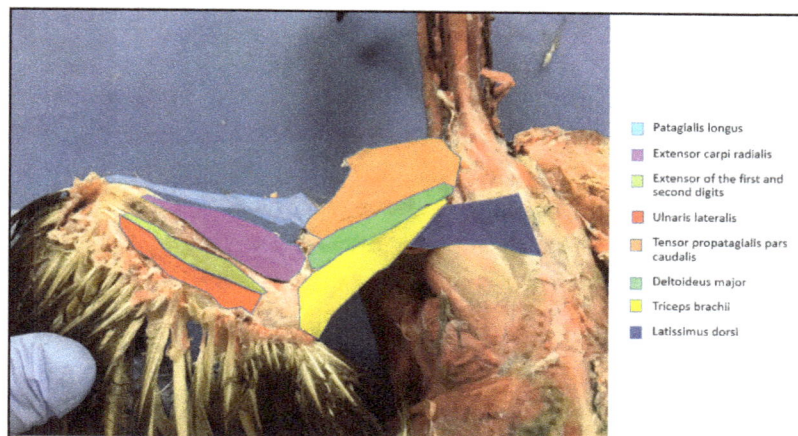

Labelled dorsal musculature of a pigeon wing.

The skin muscles help a bird in its flight by adjusting the feathers, which are attached to the skin muscle and help the bird in its flight maneuvers as well as aiding in mating rituals.

There are only a few muscles in the trunk and the tail, but they are very strong and are essential for the bird. These include the lateralis caudae and the levator caudae which control movement of the tail and the spreading of rectrices, giving the tail a larger surface area which helps keep the bird in the air as well as aiding in turning.

Muscle composition and adaptation differ by theories of muscle adaptation in whether evolution of flight came from flapping or gliding first.

Integumentary System

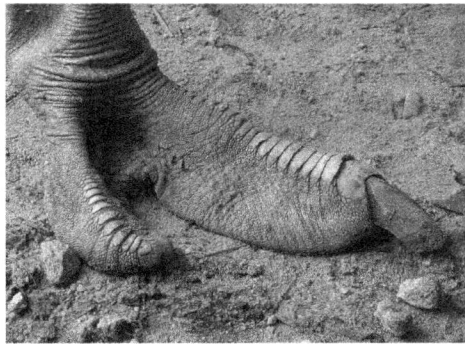

Ostrich foot integument (podotheca).

Scales

The scales of birds are composed of keratin, like beaks, claws, and spurs. They are found mainly on the toes and tarsi (lower leg of birds), usually up to the tibio-tarsal joint, but may be found further up the legs in some birds. In many of the eagles and owls the legs are feathered down to (but not including) their toes. Most bird scales do not overlap significantly, except in the cases of kingfishers and woodpeckers. The scales and scutes of birds were originally thought to be homologous to those of reptiles; however, more recent research suggests that scales in birds re-evolved after the evolution of feathers.

Bird embryos begin development with smooth skin. On the feet, the corneum, or outermost layer, of this skin may keratinize, thicken and form scales. These scales can be organized into;

- Cancella – Minute scales which are really just a thickening and hardening of the skin, criss-crossed with shallow grooves.

- Scutella – Scales that are not quite as large as scutes, such as those found on the caudal, or hind part, of the chicken metatarsus.

- Scutes – The largest scales, usually on the anterior surface of the metatarsus and dorsal surface of the toes.

The rows of scutes on the anterior of the metatarsus can be called an "acrometatarsium" or "acrotarsium".

Reticula are located on the lateral and medial surfaces (sides) of the foot and were originally thought to be separate scales. However, histological and evolutionary developmental work in this area revealed that these structures lack beta-keratin (a hallmark of reptilian scales) and are entirely composed of alpha-keratin. This, along with their unique structure, has led to the suggestion that these are actually feather buds that were arrested early in development.

Rhamphotheca and Podotheca

The bills of many waders have Herbst corpuscles which help them find prey hidden under wet sand, by detecting minute pressure differences in the water. All extant birds can move the parts of the upper jaw relative to the brain case. However this is more prominent in some birds and can be readily detected in parrots.

The region between the eye and bill on the side of a bird's head is called the lore. This region is sometimes featherless, and the skin may be tinted, as in many species of the cormorant family.

The scaly covering present on the foot of the birds is called podotheca.

Beak

The beak, bill, or rostrum is an external anatomical structure of birds which is used for eating and for grooming, manipulating objects, killing prey, fighting, probing for food, courtship and feeding young. Although beaks vary significantly in size, shape and color, they share a similar underlying structure. Two bony projections—the upper and lower mandibles—covered with a thin keratinized layer of epidermis known as the rhamphotheca. In most species, two holes known as nares lead to the respiratory system.

Respiratory System

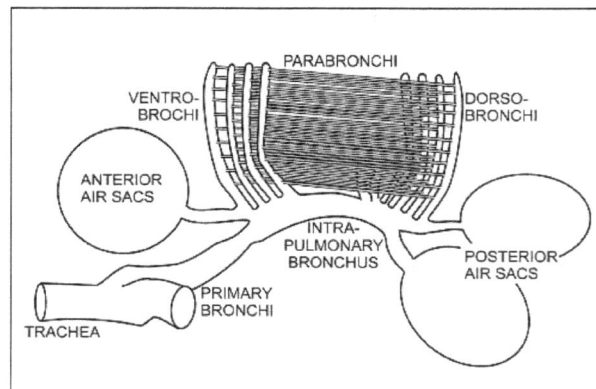

In above figure, the anatomy of bird's respiratory system, showing the relationships of the trachea, primary and intra-pulmonary bronchi, the dorso- and ventro-bronchi, with the parabronchi running between the two. The posterior and anterior air sacs are also indicated, but not to scale.

Due to the high metabolic rate required for flight, birds have a high oxygen demand. Their highly effective respiratory system helps them meet that demand.

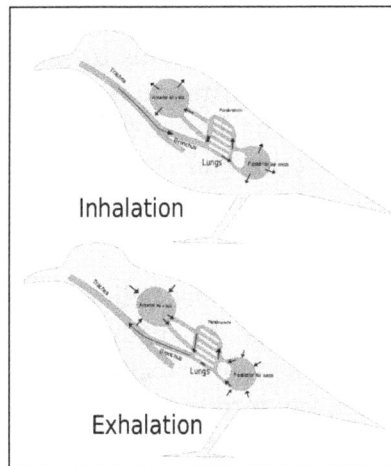

Inhalation–exhalation cycle in birds.

Although birds have lungs, theirs are fairly rigid structures that do not expand and contract as they do in mammals, reptiles and many amphibians. Instead, the structures that act as the bellows that ventilate the lungs are the air sacs, which are distributed throughout much of the birds' bodies. The airsacs move air unidirectionally through the parabronchi of the rigid lungs. Although bird lungs are smaller than those of mammals of comparable size, the air sacs account for 15% of the total body volume, whereas in mammals, the alveoli, which act as the bellows, constitute only 7% of the total body volume. The walls of the air sacs do not have a good blood supply and so do not play a direct role in gas exchange.

Birds lack a diaphragm, and therefore use their intercostal and abdominal muscles to expand and contract their entire thoraco-abdominal cavities, thus rhythmically changing the volumes of all their air sacs in unison. The active phase of respiration in birds is exhalation, requiring contraction of their muscles of respiration. Relaxation of these muscles causes inhalation.

Three distinct sets of organs perform respiration — the anterior air sacs (interclavicular, cervicals, and anterior thoracics), the lungs, and the posterior air sacs (posterior thoracics and abdominals). Typically there are nine air sacs within the system; however, that number can range between seven and twelve, depending on the species of bird. Passerines possess seven air sacs, as the clavicular air sacs may interconnect or be fused with the anterior thoracic sacs.

During inhalation, environmental air initially enters the bird through the nostrils from where it is heated, humidified, and filtered in the nasal passages and upper parts of the trachea. From there, the air enters the lower trachea and continues to just beyond the syrinx, at which point the trachea branches into two primary bronchi, going to the two lungs. The primary bronchi enter the lungs to become the intrapulmonary bronchi, which give off a set of parallel branches called ventrobronchi and, a little further on, an equivalent set of dorsobronchi. The ends of the intrapulmonary bronchi discharge air into the posterior air sacs at the caudal end of the bird. Each pair of dorso-ventrobronchi is connected by a large number of parallel microscopic air capillaries (or parabronchi) where gas exchange occurs. As the bird inhales, tracheal air flows through the intrapulmonary bronchi into the posterior air sacs, as well as into the *dorso*bronchi (but not into the ventrobronchi whose openings into the intrapulmonary bronchi were previously believed to be tightly closed during inhalation. However, more recent studies have shown that the aerodynamics of the bronchial architecture directs the inhaled air away from the openings of the ventrobronchi, into the continuation of the intrapulmonary bronchus towards the dorsobronchi and posterior air sacs). From the dorsobronchi the air flows through the parabronchi (and therefore the gas exchanger) to the ventrobronchi from where the air can only escape into the expanding anterior air sacs. So, during inhalation, both the posterior and anterior air sacs expand, the posterior air sacs filling with fresh inhaled air, while the anterior air sacs fill with "spent" (oxygen-poor) air that has just passed through the lungs.

During exhalation the intrapulmonary bronchi were believed to be tightly constricted between the region where the ventrobronchi branch off and the region where the dorsobronchi branch off. But it is now believed that more intricate aerodynamic features have the same effect. The contracting posterior air sacs can therefore only empty into the dorsobronchi. From there the fresh air from the posterior air sacs flows through the parabronchi (in the same direction as occurred during inhalation) into ventrobronchi. The air passages connecting the ventrobronchi and anterior air sacs to the intrapulmonary bronchi open up during exhalation, thus allowing oxygen-poor air from these two organs to escape via the trachea to the exterior. Oxygenated air therefore flows constantly (during the entire breathing cycle) in a single direction through the parabronchi.

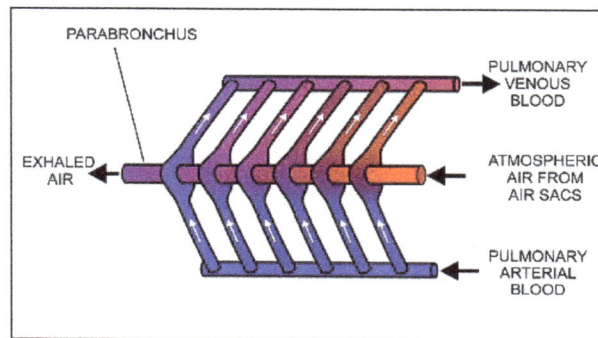

The cross-current respiratory gas exchanger in the lungs of birds. Air is forced from the air sacs unidirectionally (from right to left in the diagram) through the parabronchi. The pulmonary capillaries surround the parabronchi in the manner shown (blood flowing from below the parabronchus to above it in the diagram). Blood or air with a high oxygen content is shown in red; oxygen-poor air or blood is shown in various shades of purple-blue.

The blood flow through the bird lung is at right angles to the flow of air through the parabronchi, forming a cross-current flow exchange system. The partial pressure of oxygen in the parabronchi declines along their lengths as O_2 diffuses into the blood. The blood capillaries leaving the exchanger near the entrance of airflow take up more O_2 than do the capillaries leaving near the exit end of the parabronchi. When the contents of all capillaries mix, the final partial pressure of oxygen of the mixed pulmonary venous blood is higher than that of the exhaled air, but is nevertheless less than half that of the inhaled air, thus achieving roughly the same systemic arterial blood partial pressure of oxygen as mammals do with their bellows-type lungs.

The trachea is an area of dead space: the oxygen-poor air it contains at the end of exhalation is the first air to re-enter the posterior air sacs and lungs. In comparison to the mammalian respiratory tract, the dead space volume in a bird is, on average, 4.5 times greater than it is in mammals of the same size. Birds with long necks will inevitably have long tracheae, and must therefore take deeper breaths than mammals do to make allowances for their greater dead space volumes. In some birds (e.g. the whooper swan, *Cygnus cygnus*, the white spoonbill, *Platalea leucorodia*, the whooping crane, *Grus americana*, and the helmeted curassow, *Pauxi pauxi*) the trachea, which some cranes can be 1.5 m long, is coiled back and forth within the body, drastically increasing the dead space ventilation. The purpose of this extraordinary feature is unknown.

Air passes unidirectionally through the lungs during both exhalation and inspiration, causing, except for the oxygen-poor dead space air left in the trachea after exhalation and breathed in at the beginning of inhalation, little to no mixing of new oxygen-rich air with spent oxygen-poor air (as occurs in mammalian lungs), changing only (from oxygen-rich to oxygen-poor) as it moves (unidirectionally) through the parabronchi.

Avian lungs do not have alveoli as mammalian lungs do. Instead they contain millions of narrow passages known as parabronchi, connecting the dorsobronchi to the ventrobronchi at either ends of the lungs. Air flows anteriorly (caudal to cranial) through the parallel parabronchi. These parabronchi have honeycombed walls. The cells of the honeycomb are dead-end air vesicles, called *atria*, which project radially from the parabronchi. The *atria* are the site of gas exchange by simple diffusion. The blood flow around the parabronchi (and their atria), forms a cross-current gas exchanger.

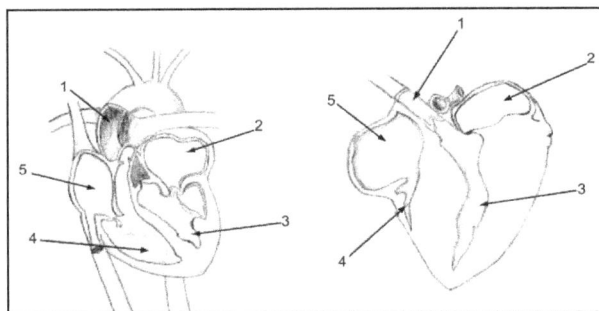

In above figure, the human heart (left) and chicken heart (right) share many similar characteristics. Avian hearts pump faster than mammalian hearts. Due to the faster heart rate, the muscles surrounding the ventricles of the chicken heart are thicker. Both hearts are labeled with the following parts: 1. Ascending Aorta 2. Left Atrium 3. Left Ventricle 4. Right Ventricle 5. Right Atrium. In chickens and others birds, the superior cava is double.

All species of birds with the exception of the penguin, have a small region of their lungs devoted to "neopulmonic parabronchi". This unorganized network of microscopic tubes branches off from the posterior air sacs, and open haphazardly into both the dorso- and ventrobronchi, as well as directly into the intrapulmonary bronchi. Unlike the parabronchi, in which the air moves unidirectionally, the air flow in the neopulmonic parabronchi is bidirectional. The neopulmonic parabronchi never make up more than 25% of the total gas exchange surface of birds.

The syrinx is the sound-producing vocal organ of birds, located at the base of a bird's trachea. As with the mammalian larynx, sound is produced by the vibration of air flowing across the organ. The syrinx enables some species of birds to produce extremely complex vocalizations, even mimicking human speech. In some songbirds, the syrinx can produce more than one sound at a time.

Circulatory System

Birds have a four-chambered heart, in common with mammals, and some reptiles (mainly the crocodilia). This adaptation allows for an efficient nutrient and oxygen transport throughout the body, providing birds with energy to fly and maintain high levels of activity. A ruby-throated hummingbird's heart beats up to 1200 times per minute (about 20 beats per second).

Digestive System

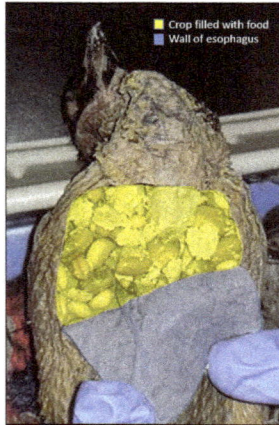

Pigeon crop containing ingested food particles is highlighted in yellow. The crop is an out-pouching of the esophagus and the wall of the esophagus is shown in blue.

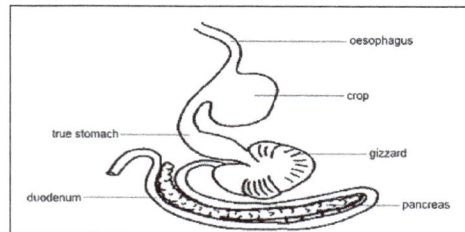

Simplified depiction of avian digestive system.

Crop

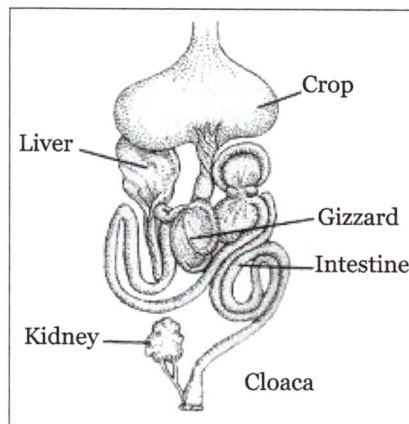

Alimentary canal of the bird exposed.

Many birds possess a muscular pouch along the esophagus called a crop. The crop functions to both soften food and regulate its flow through the system by storing it temporarily. The size and shape of the crop is quite variable among the birds. Members of the family Columbidae, such as pigeons, produce a nutritious crop milk which is fed to their young by regurgitation.

Proventriculus

The avian stomach is composed of two organs, the proventriculus and the gizzard that work together during digestion. The proventriculus is a rod shaped tube, which is found between the esophagus and the gizzard, that secretes hydrochloric acid and pepsinogen into the digestive tract. The acid converts the inactive pepsinogen into the active proteolytic enzyme, pepsin, which breaks down specific peptide bonds found in proteins, to produce a set of peptides, which are amino acid chains that are shorter than the original dietary protein. The gastric juices (hydrochloric acid and pepsinogen) are mixed with the stomach contents through the muscular contractions of the gizzard.

Gizzard

The gizzard is composed of four muscular bands that rotate and crush food by shifting the food from one area to the next within the gizzard. The gizzard of some species of herbivorous birds, like turkey and quails, contains small pieces of grit or stone called gastroliths that are swallowed by the bird to aid in the grinding process, serving the function of teeth. The use of gizzard stones is a similarity found between birds and dinosaurs, which left gastroliths as trace fossils.

Intestines

The partially digested and pulverized gizzard contents, now called a bolus, are passed into the intestine, where pancreatic and intestinal enzymes complete the digestion of the digestible food. The digestion products are then absorbed through the intestinal mucosa into the blood. The intestine ends via the large intestine in the vent or cloaca which serves as the common exit for renal and intestinal excrements as well as for the laying of eggs. However, unlike mammals, many birds do not excrete the bulky portions (roughage) of their undigested food (e.g. feathers, fur, bone fragments, and seed husks) via the cloaca, but regurgitate them as food pellets.

Drinking Behavior

There are three general ways in which birds drink: using gravity itself, sucking, and by using the tongue. Fluid is also obtained from food.

Most birds are unable to swallow by the "sucking" or "pumping" action of peristalsis in their esophagus (as humans do), and drink by repeatedly raising their heads after filling their mouths to allow the liquid to flow by gravity, a method usually described as "sipping" or "tipping up". The notable exception is the Columbidae; in fact, according to Konrad Lorenz in 1939:

> "One recognizes the order by the single behavioral characteristic, namely that in drinking the water is pumped up by peristalsis of the esophagus which occurs without exception

within the order. The only other group, however, which shows the same behavior, the Pteroclidae, is placed near the doves just by this doubtlessly very old characteristic".

Although this general rule still stands, since that time, observations have been made of a few exceptions in both directions.

In addition, specialized nectar feeders like sunbirds (Nectariniidae) and hummingbirds (Trochilidae) drink by using protrusible grooved or trough-like tongues, and parrots (Psittacidae) lap up water.

Many seabirds have glands near the eyes that allow them to drink seawater. Excess salt is eliminated from the nostrils. Many desert birds get the water that they need entirely from their food. The elimination of nitrogenous wastes as uric acid reduces the physiological demand for water, as uric acid is not very toxic and thus does not need to be diluted in as much water.

Reproductive and Urogenital Systems

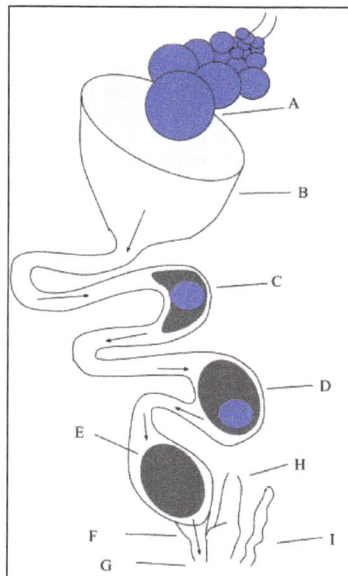

Seen here is a diagram of a female chicken reproduction system: A. Mature ovum, B. Infundibulum, C. Magnum, D. Isthmus, E. Uterus, F. Vagina, G. Cloaca, H. Large intestine, I. rudiment of right oviduct.

Fledgling

Male birds have two testes which become hundreds of times larger during the breeding season to produce sperm. The testes in birds are generally asymmetric with most birds having a larger left testis. Female birds in most families have only one functional ovary (the left one), connected to an oviduct — although two ovaries are present in the embryonic stage of each female bird. Some species of birds have two functional ovaries, and the order Apterygiformes always retain both ovaries.

Most male birds have no phallus. In the males of species without a phallus, sperm is stored in the seminal glomera within the cloacal protuberance prior to copulation. During copulation, the female moves her tail to the side and the male either mounts the female from behind or in front (as in the stitchbird), or moves very close to her. The cloacae then touch, so that the sperm can enter the female's reproductive tract. This can happen very fast, sometimes in less than half a second.

The sperm is stored in the female's sperm storage tubules for a period varying from a week to more than 100 days, depending on the species. Then, eggs will be fertilized individually as they leave the ovaries, before the shell is calcified in the oviduct. After the egg is laid by the female, the embryo continues to develop in the egg outside the female body.

A juvenile laughing gull.

Many waterfowl and some other birds, such as the ostrich and turkey, possess a phallus. This appears to be the primitive condition among birds, most birds have lost the phallus. The length is thought to be related to sperm competition in species that usually mate many times in a breeding season; sperm deposited closer to the ovaries is more likely to achieve fertilization. The longer and more complicated phalli tend to occur in waterfowl whose females have unusual anatomical features of the vagina (such as dead end sacs and clockwise coils). These vaginal structures may be used to prevent penetration by the male phallus (which coils counter-clockwise). In these species, copulation is often violent and female co-operation is not required; the female ability to prevent fertilization may allow the female to choose the father for her offspring. When not copulating, the phallus is hidden within the proctodeum compartment within the cloaca, just inside the vent.

After the eggs hatch, parents provide varying degrees of care in terms of food and protection. Precocial birds can care for themselves independently within minutes of hatching; altricial hatchlings are helpless, blind, and naked, and require extended parental care. The chicks of many ground-nesting birds such as partridges and waders are often able to run virtually immediately after hatching; such birds are referred to as nidifugous. The young of hole-nesters though, are often totally incapable of unassisted survival. The process whereby a chick acquires feathers until it can fly is called "fledging".

Some birds, such as pigeons, geese, and red-crowned cranes, remain with their mates for life and may produce offspring on a regular basis.

Kidney

Avian kidneys function in almost the same way as the more extensively studied mammalian kidney, but with a few important adaptations; while much of the anatomy remains unchanged in design, some important modifications have occurred during their evolution. A bird has paired kidneys which are connected to the lower gastrointestinal tract through the ureters. Depending on the bird species, the cortex makes up around 71-80% of the kidney's mass, while the medulla is much smaller at about 5-15% of the mass. Blood vessels and other tubes make up the remaining mass. Unique to birds is the presence of two different types of nephrons (the functional unit of the kidney) both reptilian-like nephrons located in the cortex and mammalian-like nephrons located in the medulla. Reptilian nephrons are more abundant but lack the distinctive loops of Henle seen in mammals. The urine collected by the kidney is emptied into the cloaca through the ureters and then to the colon by reverse peristalsis.

Nervous System

Birds have acute eyesight—raptors (birds of prey) have vision eight times sharper than humans—thanks to higher densities of photoreceptors in the retina (up to 1,000,000 per square mm in *Buteos*, compared to 200,000 for humans), a high number of neurons in the optic nerves, a second set of eye muscles not found in other animals, and, in some cases, an indented fovea which magnifies the central part of the visual field. Many species, including hummingbirds and albatrosses, have two foveas in each eye. Many birds can detect polarised light.

The avian ear is adapted to pick up on slight and rapid changes of pitch found in bird song. General avian tympanic membrane form is ovular and slightly conical. Morphological differences in the middle ear are observed between species. Ossicles within green finches, blackbirds, song thrushes, and house sparrows are proportionately shorter to those found in pheasants, Mallard ducks, and sea birds. In song birds, a syrinx allows the respective possessors to create intricate melodies and tones. The middle avian ear is made up of three semicircular canals, each ending in an ampulla and joining to connect with the macula sacculus and lagena, of which the cochlea, a straight short tube to the external ear, branches from.

Birds have a large brain to body mass ratio. This is reflected in the advanced and complex bird intelligence.

Immune System

The immune system of birds resembles that of other animals. Birds have both innate and adaptive immune systems. Birds are susceptible to tumours, immune deficiency and autoimmune diseases.

Bursa of Fabricius

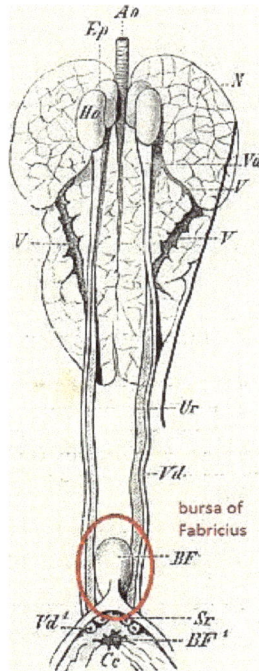

Internal view of the location of bursa of fabricius.

The bursa of fabricius, also known as the cloacal bursa, is a lymphoid organ which aids in the production of B lymphocytes during humoral immunity. The bursa of fabricius is present during juvenile stages but curls up, and in the sparrow is not visible after the sparrow reaches sexual maturity.

Anatomy

The bursa of fabricius is a circular pouch connected to the superior dorsal side of the cloaca. The bursa is composed of many folds, known as plica, which are lined by more than 10,000 follicles encompassed by connective tissue and surrounded by mesenchyme. Each follicle consists of a cortex that surrounds a medulla. The cortex houses the highly compacted B lymphocytes, whereas the medulla houses lymphocytes loosely. The medulla is separated from the lumen by the epithelium and this aids in the transport of epithelial cells into the lumen of the bursa. There are 150,000 B lymphocytes located around each follicle.

Bird Vision

Vision is the most important sense for birds, since good eyesight is essential for safe flight, and this group has a number of adaptations which give visual acuity superior to that of other vertebrate groups; a pigeon has been described as "two eyes with wings". The avian eye resembles that of a reptile, with ciliary muscles that can change the shape of the lens rapidly and to a greater extent than in the mammals. Birds have the largest eyes relative to their size in the animal kingdom, and movement is consequently limited within the eye's bony socket. In addition to the two eyelids usually found in vertebrates, it is protected by a third transparent movable membrane. The eye's internal anatomy is similar to that of other vertebrates, but has a structure, the pecten oculi, unique to birds.

Some bird groups have specific modifications to their visual system linked to their way of life. Birds of prey have a very high density of receptors and other adaptations that maximise visual acuity. The placement of their eyes gives them good binocular vision enabling accurate judgement of distances. Nocturnal species have tubular eyes, low numbers of colour detectors, but a high density of rod cells which function well in poor light. Terns, gulls and albatrosses are amongst the seabirds which have red or yellow oil droplets in the colour receptors to improve distance vision especially in hazy conditions.

Extraocular Anatomy

The eye of a bird most closely resembles that of the reptiles. Unlike the mammalian eye, it is not spherical, and the flatter shape enables more of its visual field to be in focus. A circle of bony plates, the sclerotic ring, surrounds the eye and holds it rigid, but an improvement over the reptilian eye, also found in mammals, is that the lens is pushed further forward, increasing the size of the image on the retina.

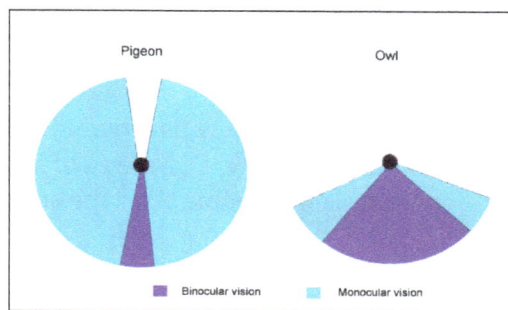

Visual fields for a pigeon and an owl.

Most birds cannot move their eyes, although there are exceptions, such as the great cormorant. Birds with eyes on the sides of their heads have a wide visual field, useful for detecting predators, while those with eyes on the front of their heads, such as owls, have binocular vision and can estimate distances when hunting. The American woodcock probably has the largest visual field of any bird, 360° in the horizontal plane, and 180° in the vertical plane.

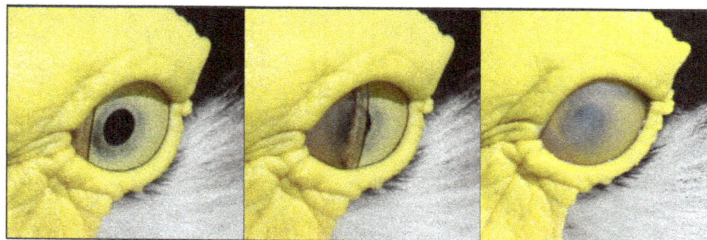

The nictitating membrane of a masked lapwing.

The eyelids of a bird are not used in blinking. Instead the eye is lubricated by the nictitating membrane, a third concealed eyelid that sweeps horizontally across the eye like a windscreen wiper. The nictitating membrane also covers the eye and acts as a contact lens in many aquatic birds when they are under water. When sleeping, the lower eyelid rises to cover the eye in most birds, with the exception of the horned owls where the upper eyelid is mobile.

The eye is also cleaned by tear secretions from the lachrymal gland and protected by an oily substance from the Harderian glands which coats the cornea and prevents dryness. The eye of a bird is larger compared to the size of the animal than for any other group of animals, although much of it is concealed in its skull. The ostrich has the largest eye of any land vertebrate, with an axial length of 50 mm (2 in), twice that of the human eye.

Bird eye size is broadly related to body mass. A study of five orders (parrots, pigeons, petrels, raptors and owls) showed that eye mass is proportional to body mass, but as expected from their habits and visual ecology, raptors and owls have relatively large eyes for their body mass.

Behavioral studies show that many avian species focus on distant objects preferentially with their lateral and monocular field of vision, and birds will orientate themselves sideways to maximise visual resolution. For a pigeon, resolution is twice as good with sideways monocular vision than forward binocular vision, whereas for humans the converse is true.

The European robin has relatively large eyes, and starts to sing early in the morning.

The performance of the eye in low light levels depends on the distance between the lens and the retina, and small birds are effectively forced to be diurnal because their eyes are not large enough to give adequate night vision. Although many species migrate at night, they often collide with even brightly lit objects like lighthouses or oil platforms. Birds of prey are diurnal because, although their eyes are large, they are optimised to give maximum spatial resolution rather than light gathering, so they also do not function well in poor light. Many birds have an asymmetry in the eye's structure which enables them to keep the horizon and a significant part of the ground in focus simultaneously. The cost of this adaptation is that they have myopia in the lower part of their visual field.

Birds with relatively large eyes compared to their body mass, such as common redstarts and European robins sing earlier at dawn than birds of the same size and smaller body mass. However, if birds have the same eye size but different body masses, the larger species sings later than the smaller. This may be because the smaller bird has to start the day earlier because of weight loss overnight. Overnight weight loss for small birds is typically 5-10% and may be over 15% on cold winter nights. In one study, robins put on more mass in their dusk feeding when nights were cold.

Nocturnal birds have eyes optimised for visual sensitivity, with large corneas relative to the eye's length, whereas diurnal birds have longer eyes relative to the corneal diameter to give greater visual acuity. Information about the activities of extinct species can be deduced from measurements of the sclerotic ring and orbit depth. For the latter measurement to be made, the fossil must have retained its three-dimensional shape, so activity pattern cannot be determined with confidence from flattened specimens like *Archaeopteryx*, which has a complete sclerotic ring but no orbit depth measurement.

Anatomy of the Eye

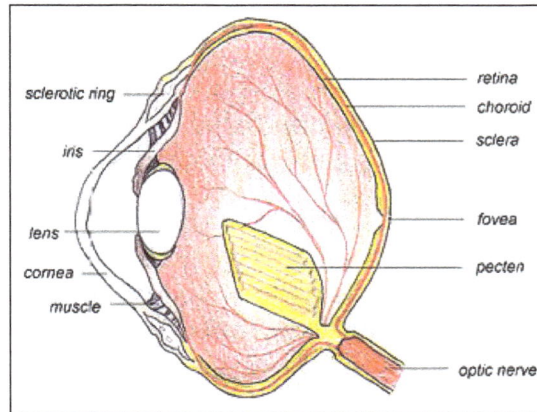

Anatomy of the avian eye.

The main structures of the bird eye are similar to those of other vertebrates. The outer layer of the eye consists of the transparent cornea at the front, and two layers of sclera — a tough white collagen fibre layer which surrounds the rest of the eye and supports and protects the eye as a whole. The eye is divided internally by the lens into two main segments: the anterior segment and the posterior segment. The anterior chamber is filled with a watery fluid called the aqueous humour, and the posterior chamber contains the vitreous humour, a clear jelly-like substance.

The lens is a transparent convex or 'lens' shaped body with a harder outer layer and a softer inner layer. It focuses the light on the retina. The shape of the lens can be altered by ciliary muscles which are directly attached to the lens capsule by means of the zonular fibres. In addition to these muscles, some birds also have a second set, Crampton's muscles, that can change the shape of the cornea, thus giving birds a greater range of accommodation than is possible for mammals. This accommodation can be rapid in some diving water birds such as in the mergansers. The iris is a coloured muscularly operated diaphragm in front of the lens which controls the amount of light entering the eye. At the centre of the iris is the pupil, the variable circular area through which the light passes into the eye.

Hummingbirds are amongst the many birds with two foveae.

The retina is a relatively smooth curved multi-layered structure containing the photosensitive rod and cone cells with the associated neurons and blood vessels. The density of the photoreceptors is critical in determining the maximum attainable visual acuity. Humans have about 200,000 receptors per mm^2, but the house sparrow has 400,000 and the common buzzard 1,000,000. The photoreceptors are not all individually connected to the optic nerve, and the ratio of nerve ganglia to receptors is important in determining resolution. This is very high for birds; the white wagtail has 100,000 ganglion cells to 120,000 photoreceptors.

Rods are more sensitive to light, but give no colour information, whereas the less sensitive cones enable colour vision. In diurnal birds, 80% of the receptors may be cones (90% in some swifts) whereas nocturnal owls have almost all rods. As with other vertebrates except placental mammals, some of the cones may be double cones. These can amount to 50% of all cones in some species.

Towards the centre of the retina is the fovea (or the less specialised, area centralis) which has a greater density of receptors and is the area of greatest forward visual acuity, i.e. sharpest, clearest detection of objects. In 54% of birds, including birds of prey, kingfishers, hummingbirds and swallows, there is second fovea for enhanced sideways viewing. The optic nerve is a bundle of nerve fibres which carry messages from the eye to the relevant parts of the brain. Like mammals, birds have a small blind spot without photoreceptors at the optic disc, under which the optic nerve and blood vessels join the eye.

The pecten is a poorly understood body consisting of folded tissue which projects from the retina. It is well supplied with blood vessels and appears to keep the retina supplied with nutrients, and may also shade the retina from dazzling light or aid in detecting moving objects. Pecten oculi is abundantly filled with melanin granules which have been proposed to absorb stray light entering the bird eye to reduce background glare. Slight warming of pecten oculi due to absorption of light by melanin granules has been proposed to enhance metabolic rate of pecten. This is suggested to help increase secretion of nutrients into the vitreous body, eventually to be absorbed by the avascular retina of birds for improved nutrition. Extra-high enzymic activity of alkaline phosphatase in pecten oculi has been proposed to support high secretory activity of pecten to supplement nutrition of the retina.

The choroid is a layer situated behind the retina which contains many small arteries and veins. These provide arterial blood to the retina and drain venous blood. The choroid contains melanin, a pigment which gives the inner eye its dark colour, helping to prevent disruptive reflections.

Light Perception

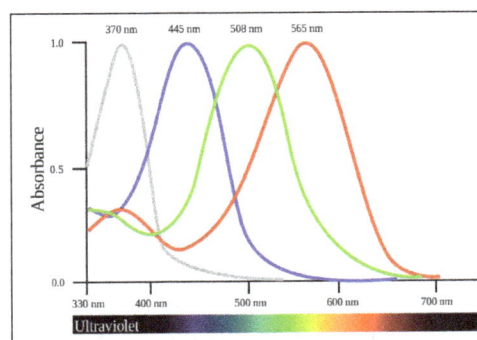

The four pigments in estrildid finches' cones extend the range of colour vision into the ultraviolet.

There are two sorts of light receptors in a bird's eye, rods and cones. Rods, which contain the visual pigment rhodopsin are better for night vision because they are sensitive to small quantities of light. Cones detect specific colours (or wavelengths) of light, so they are more important to colour-orientated animals such as birds. Most birds are tetrachromatic, possessing four types of cone cells each with a distinctive maximal absorption peak. In some birds, the maximal absorption peak of the cone cell responsible for the shortest wavelength extends to the ultraviolet (UV) range, making them UV-sensitive. In addition to that, the cones at the bird's retina are arranged in a characteristic form of spatial distribution, known as hyperuniform distribution, which maximizes its light and color absorption. This form of spatial distributions are only observed as a result of some optimization process, which in this case can be described in terms of birds' evolutionary history.

The four spectrally distinct cone pigments are derived from the protein opsin, linked to a small molecule called retinal, which is closely related to vitamin A. When the pigment absorbs light the retinal changes shape and alters the membrane potential of the cone cell affecting neurons in the ganglia layer of the retina. Each neuron in the ganglion layer may process information from a number of photoreceptor cells, and may in turn trigger a nerve impulse to relay information along the optic nerve for further processing in specialised visual centres in the brain. The more intense a light, the more photons are absorbed by the visual pigments; the greater the excitation of each cone, and the brighter the light appears.

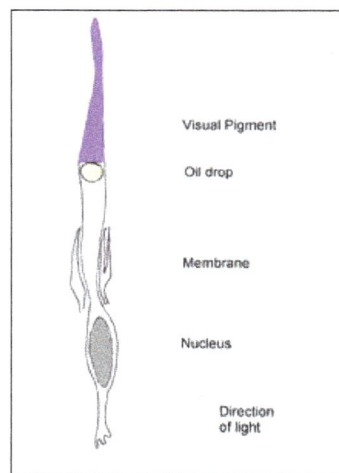

Diagram of a bird cone cell.

By far the most abundant cone pigment in every bird species examined is the long-wavelength form of iodopsin, which absorbs at wavelengths near 570 nm. This is roughly the spectral region occupied by the red- and green-sensitive pigments in the primate retina, and this visual pigment dominates the colour sensitivity of birds. In penguins, this pigment appears to have shifted its absorption peak to 543 nm, presumably an adaptation to a blue aquatic environment.

The information conveyed by a single cone is limited: by itself, the cell cannot tell the brain which wavelength of light caused its excitation. A visual pigment may absorb two wavelengths equally, but even though their photons are of different energies, the cone cannot tell them apart, because they both cause the retinal to change shape and thus trigger the same impulse. For the brain to see colour, it must compare the responses of two or more classes of cones containing different visual pigments, so the four pigments in birds give increased discrimination.

Each cone of a bird or reptile contains a coloured oil droplet; these no longer exist in mammals. The droplets, which contain high concentrations of carotenoids, are placed so that light passes through them before reaching the visual pigment. They act as filters, removing some wavelengths and narrowing the absorption spectra of the pigments. This reduces the response overlap between pigments and increases the number of colours that a bird can discern. Six types of cone oil droplets have been identified; five of these have carotenoid mixtures that absorb at different wavelengths and intensities, and the sixth type has no pigments. The cone pigments with the lowest maximal absorption peak, including those that are UV-sensitive, possess the 'clear' or 'transparent' type of oil droplets with little spectral tuning effect.

The colours and distributions of retinal oil droplets vary considerably among species, and is more dependent on the ecological niche utilised (hunter, fisher, herbivore) than genetic relationships. As examples, diurnal hunters like the barn swallow and birds of prey have few coloured droplets, whereas the surface fishing common tern has a large number of red and yellow droplets in the dorsal retina. The evidence suggests that oil droplets respond to natural selection faster than the cone's visual pigments. Even within the range of wavelengths that are visible to humans, passerine birds can detect colour differences that humans do not register. This finer discrimination, together with the ability to see ultraviolet light, means that many species show sexual dichromatism that is visible to birds but not humans.

Migratory songbirds use the Earth's magnetic field, stars, the Sun, and other unknown cues to determine their migratory direction. An American study suggested that migratory Savannah sparrows used polarised light from an area of sky near the horizon to recalibrate their magnetic navigation system at both sunrise and sunset. This suggested that skylight polarisation patterns are the primary calibration reference for all migratory songbirds. However, it appears that birds may be responding to secondary indicators of the angle of polarisation, and may not be actually capable of directly detecting polarisation direction in the absence of these cues.

Ultraviolet Sensitivity

The common kestrel, like other raptorial birds, have a very low sensitivity to UV light.

Many species of birds are tetrachromatic, with dedicated cone cells for perceiving wavelengths in the ultraviolet and violet regions of the light spectrum. These cells contain a combination of short wave sensitive (SWS1) opsins, SWS1-like opsins (SWS2), and long-wave filtering carotenoid pigments for selectively filtering and receiving light between 300 and 400 nm. There are two types of short wave color vision in birds: violet sensitive (VS) and ultraviolet sensitive (UVS). Single nucleotide substitutions in the SWS1 opsin sequence are responsible blue-shifting the spectral sensitivity of the opsin from violet sensitive (λ_{max} = 400) to ultraviolet sensitive (λ_{max} = 310-360). This is the proposed evolutionary mechanism by which ultraviolet vision originally arose. The major clades of birds that have UVS vision are Palaeognathae (ratites and tinamous), Charadriiformes (shorebirds, gulls, and alcids), Trogoniformes (trogons), Psittaciformes (parrots), and Passeriformes (perching birds, representing more than half of all avian species).

UVS vision can be useful for courtship. Birds that do not exhibit sexual dichromatism in visible wavelengths are sometimes distinguished by the presence of ultraviolet reflective patches on their feathers. Male blue tits have an ultraviolet reflective crown patch which is displayed in courtship by posturing and raising of their nape feathers. Male blue grosbeaks with the brightest and most UV-shifted blue in their plumage are larger, hold the most extensive territories with abundant prey, and feed their offspring more frequently than other males.

The bill's appearance is important in the interactions of the blackbird. Although the UV component seems unimportant in interactions between territory-holding males, where the degree of orange is the main factor, the female responds more strongly to males with bills with good UV-reflectiveness.

UVS is also demonstrated to serve functions in foraging, prey identification, and frugivory. Similar advantages afforded to trichromatic primates over dichromatic primates in frugivory are generally considered to exist in birds. The waxy surfaces of many fruits and berries reflect UV light that advertise their presence to UVS birds. Common kestrels are able to locate the trails of voles with vision; these small rodents lay scent trails of urine and feces that reflect UV light, making them visible to the kestrels. However, this view has been challenged by the finding of low UV sensitivity in raptors and weak UV reflection of mammal urine.

While tetrachromatic vision is not exclusive to birds (insects, reptiles, and crustaceans are also sensitive to short wavelengths), some predators of UVS birds cannot see ultraviolet light. This raises the possibility that ultraviolet vision gives birds a channel in which they can privately signal, thereby remaining inconspicuous to predators. However, recent evidence does not appear to support this hypothesis.

Perception

Contrast Sensitivity

Contrast (or more precisely Michelson-contrast) is defined as the difference in luminance between two stimulus areas, divided by the sum of luminance of the two. Contrast sensitivity is the inverse of the smallest contrast that can be detected; a contrast sensitivity of 100 means that the smallest contrast that can be detected is 1%. Birds have comparably lower contrast sensitivity than mammals. Humans have been shown to detect contrasts as low as 0.5-1% whereas most birds tested require ca. 10% contrast to show a behavioral response. A contrast sensitivity function describes an animal's ability to detect the contrast of grating patterns of different spatial frequency (i.e. different detail).

For stationary viewing experiments the contrast sensitivity is highest at a medium spatial frequency and lower for higher and lower spatial frequencies.

Movement

A red kite flying at a bird feeding station in Scotland.

Birds can resolve rapid movements better than humans, for whom flickering at a rate greater than 50 light pulse cycles per second appears as continuous movement. Humans cannot therefore distinguish individual flashes of a fluorescent light bulb oscillating at 60 light pulse cycles per second, but budgerigars and chickens have flicker or light pulse cycles per second thresholds of more than 100 light pulse cycles per second. A Cooper's hawk can pursue agile prey through woodland and avoid branches and other objects at high speed; to humans such a chase would appear as a blur.

Birds can also detect slow moving objects. The movement of the sun and the constellations across the sky is imperceptible to humans, but detected by birds. The ability to detect these movements allows migrating birds to properly orient themselves.

To obtain steady images while flying or when perched on a swaying branch, birds hold the head as steady as possible with compensating reflexes. Maintaining a steady image is especially relevant for birds of prey. Because the image can be centered on the deep fovea of only one eye at a time, most falcons when diving use a spiral path to approach their prey after they have locked on to a target individual. The alternative of turning the head for a better view slows down the dive by increasing drag while spiralling does not reduce speeds significantly.

Edges and Shapes

When an object is partially blocked by another, humans unconsciously tend to make up for it and complete the shapes. It has however been demonstrated that pigeons do not complete occluded shapes. A study based on altering the grey level of a perch that was coloured differently from the background showed that budgerigars do not detect edges based on colours.

Magnetic Fields

The perception of magnetic fields by migratory birds has been suggested to be light dependent. Birds move their head to detect the orientation of the magnetic field, and studies on the neural pathways have suggested that birds may be able to "see" the magnetic fields. The right eye of a migratory bird contains photoreceptive proteins called cryptochromes. Light excites these molecules to produce unpaired electrons that interact with the Earth's magnetic field, thus providing directional information.

Variations across Bird Groups

Diurnal Birds of Prey

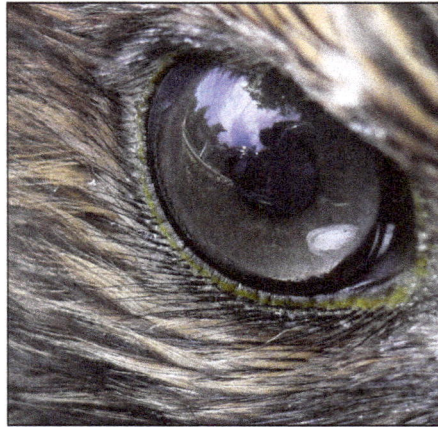

"Hawk-eyed" is a byword for visual acuity.

The visual ability of birds of prey is legendary, and the keenness of their eyesight is due to a variety of factors. Raptors have large eyes for their size, 1.4 times greater than the average for birds of the same weight, and the eye is tube-shaped to produce a larger retinal image. The resolving power of an eye depends both on the optics, large eyes with large apertures suffers less from diffraction and can have larger retinal images due to a long focal length, and on the density of receptor spacing. The retina has a large number of receptors per square millimeter, which determines the degree of visual acuity. The more receptors an animal has, the higher its ability to distinguish individual objects at a distance, especially when, as in raptors, each receptor is typically attached to a single ganglion. Many raptors have foveas with far more rods and cones than the human fovea (65,000/ mm^2 in American kestrel, 38,000 in humans) and this provides these birds with spectacular long distance vision. It is proposed that the shape of the deep central fovea of raptors can create a tele-photo optical system, increasing the size of the retinal image in the fovea and thereby increasing the spatial resolution. Behavioral studies show that some large eyed raptors (Wedge-tailed eagle, Old world vultures) and have ca 2 times higher spatial resolution than humans, but many medium and small sized raptors have comparable or lower spatial resolution.

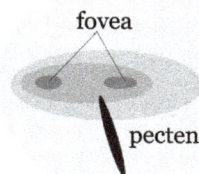

Each retina of the black-chested buzzard-eagle has two foveae.

The forward-facing eyes of a bird of prey give binocular vision, which is assisted by a double fovea. The raptor's adaptations for optimum visual resolution (an American kestrel can see a 2–mm in-sect from the top of an 18–m tree) has a disadvantage in that its vision is poor in low light level, and it must roost at night. Raptors may have to pursue mobile prey in the lower part of their visual field, and therefore do not have the lower field myopia adaptation demonstrated by many other birds. Scavenging birds like vultures do not need such sharp vision, so a condor has only a single fovea with about 35,000 receptors mm². Vultures, however have high physiological activity of many important enzymes to suit their distant clarity of vision. Southern Caracara also only have a single fovea as this species forages on the ground for carrion and insects. However, they do have a higher degree of binocular overlap than other falcons, potentially to enable the caracara to manipulate objects, such as rocks, whilst foraging.

Like other birds investigated raptors do also have coloured oil droplets in their cones. The generally brown, grey and white plumage of this group, and the absence of colour displays in courtship suggests that colour is relatively unimportant to these birds.

In most raptors a prominent eye ridge and its feathers extends above and in front of the eye. This "eyebrow" gives birds of prey their distinctive stare. The ridge physically protects the eye from wind, dust, and debris and shields it from excessive glare. The osprey lacks this ridge, although the arrangement of the feathers above its eyes serves a similar function; it also possesses dark feathers in front of the eye which probably serve to reduce the glare from the water surface when the bird is hunting for its staple diet of fish.

Nocturnal Birds

A powerful owl photographed at night showing reflective tapeta lucida.

Owls have very large eyes for their size, 2.2 times greater than the average for birds of the same weight, and positioned at the front of the head. The eyes have a field overlap of 50–70%, giving better binocular vision than for diurnal birds of prey (overlap 30–50%). The tawny owl's retina has about 56,000 light-sensitive rods per square millimetre (36 million per square inch); although earlier claims that it could see in the infrared part of the spectrum have been dismissed.

Adaptations to night vision include the large size of the eye, its tubular shape, large numbers of closely packed retinal rods, and an absence of cones, since cone cells are not sensitive enough for a low-photon nighttime environment. There are few coloured oil droplets, which would reduce the light intensity, but the retina contains a reflective layer, the tapetum lucidum. This increases the amount of light each photosensitive cell receives, allowing the bird to see better in low light conditions. Owls normally have only one fovea, and that is poorly developed except in diurnal hunters like the short-eared owl.

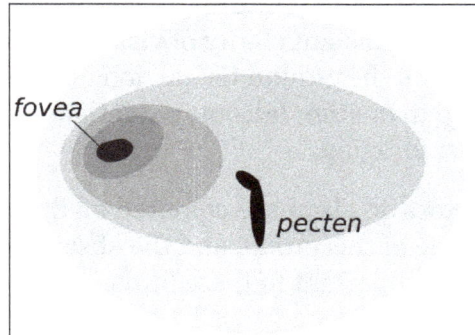

Each owl's retina has a single fovea.

Besides owls, bat hawks, frogmouths and nightjars also display good night vision. Some bird species nest deep in cave systems which are too dark for vision, and find their way to the nest with a simple form of echolocation. The oilbird is the only nocturnal bird to echolocate, but several *Aerodramus* swiftlets also utilise this technique, with one species, Atiu swiftlet, also using echolocation outside its caves.

Water Birds

Terns have coloured oil droplets in the cones of the eye to improve distance vision.

Seabirds such as terns and gulls that feed at the surface or plunge for food have red oil droplets in the cones of their retinas. This improves contrast and sharpens distance vision, especially in hazy conditions. Birds that have to look through an air/water interface have more deeply coloured carotenoid pigments in the oil droplets than other species.

This helps them to locate shoals of fish, although it is uncertain whether they are sighting the phytoplankton on which the fish feed, or other feeding birds.

Birds that fish by stealth from above the water have to correct for refraction particularly when the fish are observed at an angle. Reef herons and little egrets appear to be able to make the corrections needed when capturing fish and are more successful in catching fish when strikes are made at an acute angle and this higher success may be due to the inability of the fish to detect their predators. Other studies indicate that egrets work within a preferred angle of strike and that the probability of misses increase when the angle becomes too far from the vertical leading to an increased difference between the apparent and real depth of prey.

Birds that pursue fish under water like auks and divers have far fewer red oil droplets, but they have special flexible lenses and use the nictitating membrane as an additional lens. This allows greater optical accommodation for good vision in air and water. Cormorants have a greater range of visual accommodation, at 50 dioptres, than any other bird, but the kingfishers are considered to have the best all-round (air and water) vision.

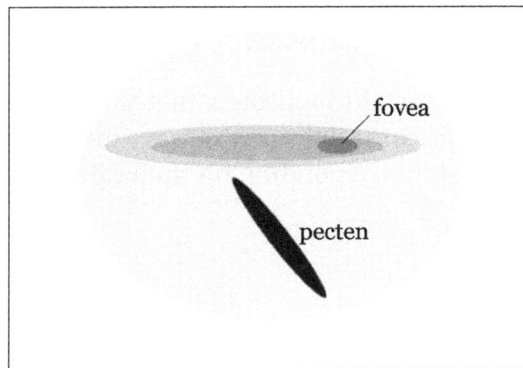

Each retina of the Manx shearwater has one fovea and an elongated strip of high photo receptor density.

Tubenosed seabirds, which come ashore only to breed and spend most of their life wandering close to the surface of the oceans, have a long narrow area of visual sensitivity on the retina. This region, the *area giganto cellularis*, has been found in the Manx shearwater, Kerguelen petrel, great shearwater, broad-billed prion and common diving-petrel. It is characterised by the presence of ganglion cells which are regularly arrayed and larger than those found in the rest of the retina, and morphologically appear similar to the cells of the retina in cats. The location and cellular morphology of this novel area suggests a function in the detection of items in a small binocular field projecting below and around the bill. It is not concerned primarily with high spatial resolution, but may assist in the detection of prey near the sea surface as a bird flies low over it.

The Manx shearwater, like many other seabirds, visits its breeding colonies at night to reduce the chances of attack by aerial predators. Two aspects of its optical structure suggest that the eye of this species is adapted to vision at night. In the shearwater's eyes the lens does most of the bending of light necessary to produce a focused image on the retina. The cornea, the outer covering of the eye, is relative flat and so of low refractive power. In a diurnal bird like the pigeon, the reverse is true; the cornea is highly curved and is the principal refractive component. The ratio of refraction by the lens to that by the cornea is 1.6 for the shearwater and 0.4 for the pigeon; the figure for the shearwater is consistent with that for a range of nocturnal birds and mammals.

The shorter focal length of shearwater eyes give them a smaller, but brighter, image than is the case for pigeons, so the latter has sharper daytime vision. Although the Manx shearwater has adaptations for night vision, the effect is small, and it is likely that these birds also use smell and hearing to locate their nests.

It used to be thought that penguins were far-sighted on land. Although the cornea is flat and adapted to swimming underwater, the lens is very strong and can compensate for the reduced corneal focusing when out of water. Almost the opposite solution is used by the hooded merganser which can bulge part of the lens through the iris when submerged.

Bird Physiology

Bird physiology is unique and closely linked to the energetic demands of flying. Birds have high metabolisms and body temperatures. Their respiratory system is highly adapted to allow efficient oxygen delivery into the body during flight. A large heart pumps the circulation system to deliver oxygen and important compound around the body.

Birds are heated by the heat generated through their metabolism. Their high body temperature gives birds faster reflexes and muscle contractions which are important attributes for flying. Many birds have specialized digestive systems depending on their specific diets and the toxins they are required to digest.

Bird Respiration

The respiratory system of birds is significantly different to mammals, mostly to account for the physical demands of flying. A bird's respiratory system is made up of the nostrils, windpipe, bronchi, two small lungs and a network of interconnected air sacs. The process of air circulation through a bird's body goes in one direction over two complete breathing cycles. This process can be separated into four simple steps. The majority of birds inhale air through their nostrils which travels through the windpipe. The windpipe splits into two primary bronchi which then divide further into multiple smaller stems.

Step 1: In the first step of the respiration cycle, birds inhale air through nostrils in their beaks which travels through the windpipe, primary bronchi and ends up in the posterior air sacs.

Step 2: During the second step, the bird exhales and the inhale air from the previous breath moves from the posterior air sac into the lungs. At this point the oxygen in the air is swapped with CO_2 in the blood.

Step 3: In step three the bird inhales for the second time and the air in the lungs moves into the anterior air sacs located towards the front of the body while the newly inhaled air is transported to the posterior air sacs (step 1).

Step 4: For the final step, the bird exhales for the second time and the air is transported from the anterior air sacs out through its beak via the windpipe.

Circulation

The circulatory system of birds must be very efficient in order to keep up with demands from the respiratory system and the bird's metabolism. The blood is responsible for transporting important compounds (such as glucose, fats and waste products) through the body. Large volumes of blood are pumped around the body of birds in order to maintain the delivery of such compounds.

The blood is also responsible for transporting oxygen and CO_2 around the body and must keep up with the respiratory system in order to do so. Because the demand for blood is higher in birds than it is in mammals, birds typically have larger hearts than mammals of similar sizes. They also have much higher heart rates with resting heart rates generally sitting between 150-350 beats per minutes for a medium sized bird. In small birds, heart rates can reach in excess of 1200 beats per minute. The majority of blood is sent to major organs such as the kidneys, liver, lungs and intestines.

Metabolism

The metabolism is the set of chemical reactions that keep an organism alive and is an important part of a bird's physiology. These reactions require energy and an individual's metabolism is the total amount of energy required to sustain this complete set of chemical reactions. If we consider all organisms and more specifically just animals, birds have high metabolisms relative to most living things. They are active, warm-blooded animals so they require large amounts of energy to maintain their active lifestyle and their body temperatures.

Their metabolisms are highest whilst flying when they require the most energy and lowest while they are sleeping. Metabolism in birds is closely linked to temperature regulation. Small birds have the highest metabolisms relative to their size because their small bodies lose heat faster than larger birds so their metabolisms must work harder in order to maintain their internal body temperatures.

Digestion

Bird physiology of the digestive system in birds can be very specialized. Birds can have specialized digestive systems for different diets and they can change significantly with changes in seasons. They have sacs on the sides of their large intestines called 'ceca' which helps birds to digest plant material. Some birds are able to digest waxes and parrots eat clay to help them digest toxic compounds in fruits and seeds. Hummingbirds are able to extract 99% of the energy from the nectar they feed on.

Temperature Regulation

Birds maintain their body temperatures by producing heat through their metabolism rather than relying on heat from the environment. Their feathers also play an important role in regulating their temperatures by controlling how much heat leaves the body.

A bird's body temperature is usually maintained within the high range of 40-42 °C and this allows birds to be very active but means they are required to eat a lot of food. A bird will consume around 20 times as much food as a reptile of a similar size. Birds also regulate their body temperatures by shivering when they are cold, panting or fluttering when they are hot, and increasing or reducing blood flow to their feet. Temperatures over 46 °C are fatal.

Fish Anatomy

Fish anatomy is the study of the form or morphology of fishes. It can be contrasted with fish physiology, which is the study of how the component parts of fish function together in the living fish. In practice, fish anatomy and fish physiology complement each other, the former dealing with the structure of a fish, its organs or component parts and how they are put together, such as might be observed on the dissecting table or under the microscope, and the latter dealing with how those components function together in living fish.

The anatomy of fish is often shaped by the physical characteristics of water, the medium in which fish live. Water is much denser than air, holds a relatively small amount of dissolved oxygen, and absorbs more light than air does. The body of a fish is divided into a head, trunk and tail, although the divisions between the three are not always externally visible. The skeleton, which forms the support structure inside the fish, is either made of cartilage, in cartilaginous fish, or bone in bony fish. The main skeletal element is the vertebral column, composed of articulating vertebrae which are lightweight yet strong. The ribs attach to the spine and there are no limbs or limb girdles. The main external features of the fish, the fins, are composed of either bony or soft spines called rays which, with the exception of the caudal fins, have no direct connection with the spine. They are supported by the muscles which compose the main part of the trunk. The heart has two chambers and pumps the blood through the respiratory surfaces of the gills and on round the body in a single circulatory loop. The eyes are adapted for seeing underwater and have only local vision. There is an inner ear but no external or middle ear. Low frequency vibrations are detected by the lateral line system of sense organs that run along the length of the sides of fish, and these respond to nearby movements and to changes in water pressure.

Sharks and rays are basal fish with numerous primitive anatomical features similar to those of ancient fish, including skeletons composed of cartilage. Their bodies tend to be dorso-ventrally flattened, they usually have five pairs of gill slits and a large mouth set on the underside of the head. The dermis is covered with separate dermal placoid scales. They have a cloaca into which the urinary and genital passages open, but not a swim bladder. Cartilaginous fish produce a small number of large, yolky eggs. Some species are ovoviviparous and the young develop internally but others are oviparous and the larvae develop externally in egg cases.

The bony fish lineage shows more derived anatomical traits, often with major evolutionary changes from the features of ancient fish. They have a bony skeleton, are generally laterally flattened, have five pairs of gills protected by an operculum, and a mouth at or near the tip of the snout. The dermis is covered with overlapping scales. Bony fish have a swim bladder which helps them maintain a constant depth in the water column, but not a cloaca. They mostly spawn a large number of small eggs with little yolk which they broadcast into the water column.

Body

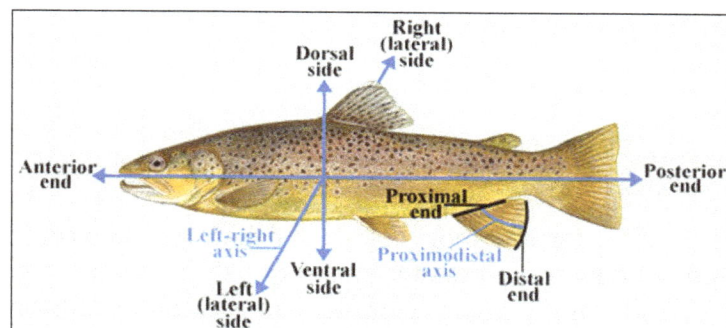

Anatomical directions and axes.

In many respects fish anatomy is different from humans and mammals, yet it shares the same basic vertebrate body plan from which all vertebrates have evolved: a notochord, rudimentary vertebrae, and a well-defined head and tail.

Fish have a variety of different body plans. At the broadest level their body is divided into head, trunk, and tail, although the divisions are not always externally visible. The body is often fusiform, a streamlined body plan often found in fast-moving fish. They may also be filiform (eel-shaped) or vermiform (worm-shaped). Also, fish are often either compressed (laterally thin) or depressed (dorso-ventrally flat).

Skeleton

Skeleton of a bony fish.

There are two different skeletal types: the exoskeleton, which is the stable outer shell of an organism, and the endoskeleton, which forms the support structure inside the body. The skeleton of the fish is made of either cartilage (cartilaginous fishes) or bone (bony fishes). The main features of the fish, the fins, are bony fin rays and, except for the caudal fin, have no direct connection with the spine. They are supported only by the muscles. The ribs attach to the spine.

Bones are rigid organs that form part of the endoskeleton of vertebrates. They function to move, support, and protect the various organs of the body, produce red and white blood cells and store minerals. Bone tissue is a type of dense connective tissue. Bones come in a variety of shapes and have a complex internal and external structure. They are lightweight, yet strong and hard, in addition to fulfilling their many other functions.

Vertebrae

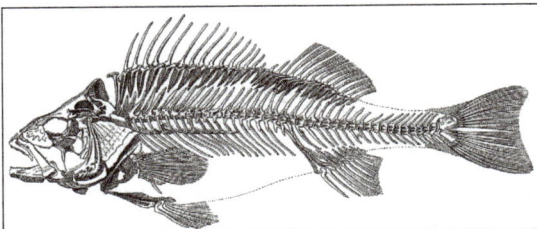

Skeletal structure of a bass showing the vertebral column running from the head to the tail.

Skeletal structure of an Atlantic cod.

The X-ray tetra *(Pristella maxillaris)* has a visible backbone. The spinal cord is housed within its backbone

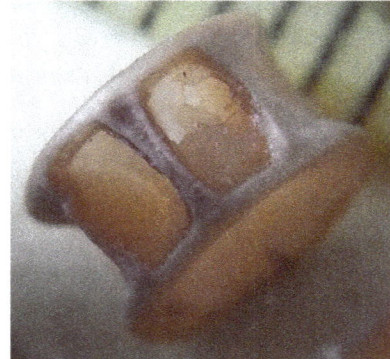

A vertebra (diameter 5 mm) of a small ray-finned fish

Fish are vertebrates. All vertebrates are built along the basic chordate body plan: a stiff rod running through the length of the animal (vertebral column or notochord), with a hollow tube of nervous tissue (the spinal cord) above it and the gastrointestinal tract below. In all vertebrates, the mouth is found at, or right below, the anterior end of the animal, while the anus opens to the exterior before the end of the body. The remaining part of the body continuing aft of the anus forms a tail with vertebrae and spinal cord, but no gut.

The defining characteristic of a vertebrate is the vertebral column, in which the notochord (a stiff rod of uniform composition) found in all chordates has been replaced by a segmented series of stiffer elements (vertebrae) separated by mobile joints (intervertebral discs, derived embryonically and evolutionarily from the notochord). However, a few fish have secondarily lost this anatomy, retaining the notochord into adulthood, such as the sturgeon.

The vertebral column consists of a centrum (the central body or spine of the vertebra), vertebral arches which protrude from the top and bottom of the centrum, and various processes which project from the centrum or arches. An arch extending from the top of the centrum is called a neural arch, while the hemal arch or chevron is found underneath the centrum in the caudal (tail) vertebrae of fish. The centrum of a fish is usually concave at each end (amphicoelous), which limits the motion of the fish. This can be contrasted with the centrum of a mammal, which is flat at each end (acoelous), shaped in a manner that can support and distribute compressive forces.

The vertebrae of lobe-finned fishes consist of three discrete bony elements. The vertebral arch surrounds the spinal cord, and is of broadly similar form to that found in most other vertebrates. Just beneath the arch lies a small plate-like *pleurocentrum*, which protects the upper surface of the notochord, and below that, a larger arch-shaped *intercentrum* to protect the lower border. Both of these structures are embedded within a single cylindrical mass of cartilage. A similar arrangement was found in primitive tetrapods, but, in the evolutionary line that led to reptiles (and hence, also to mammals and birds), the intercentrum became partially or wholly replaced by an enlarged pleurocentrum, which in turn became the bony vertebral body.

In most ray-finned fishes, including all teleosts, these two structures are fused with, and embedded within, a solid piece of bone superficially resembling the vertebral body of mammals. In living amphibians, there is simply a cylindrical piece of bone below the vertebral arch, with no trace of the separate elements present in the early tetrapods.

In cartilagenous fish, such as sharks, the vertebrae consist of two cartilagenous tubes. The upper tube is formed from the vertebral arches, but also includes additional cartilagenous structures filling in the gaps between the vertebrae, and so enclosing the spinal cord in an essentially continuous sheath. The lower tube surrounds the notochord, and has a complex structure, often including multiple layers of calcification.

Lampreys have vertebral arches, but nothing resembling the vertebral bodies found in all higher vertebrates. Even the arches are discontinuous, consisting of separate pieces of arch-shaped cartilage around the spinal cord in most parts of the body, changing to long strips of cartilage above and below in the tail region. Hagfishes lack a true vertebral column, and are therefore not properly considered vertebrates, but a few tiny neural arches are present in the tail. Hagfishes do, however, possess a cranium. For this reason, the vertebrate subphylum is sometimes referred to as "Craniata" when discussing morphology. Molecular analysis since 1992 has suggested that the hagfishes are most closely related to lampreys, and so also are vertebrates in a monophyletic sense. Others consider them a sister group of vertebrates in the common taxon of Craniata.

Head

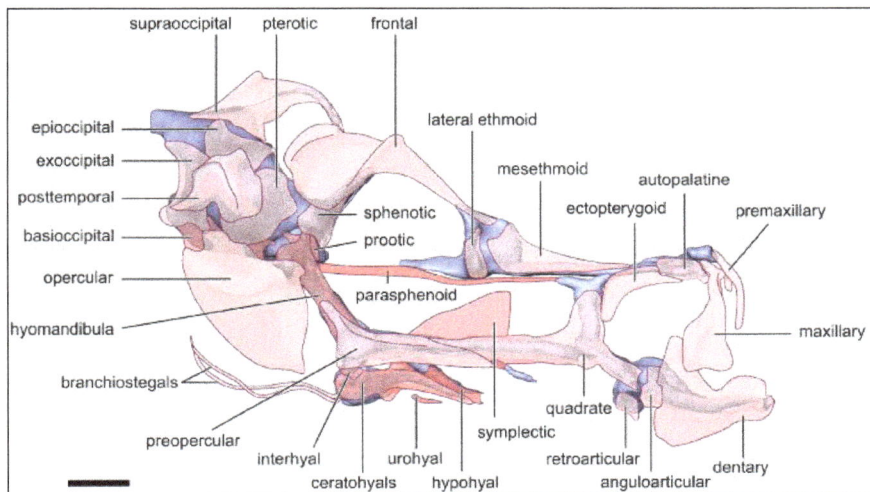

Skull bones as they appear in a seahorse.

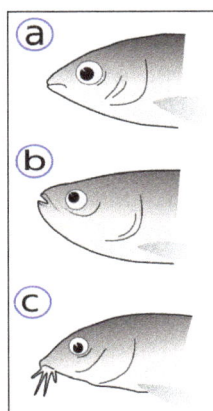

Positions of fish mouths: (a) Terminal, (b) Superior, (c) Subterminal.

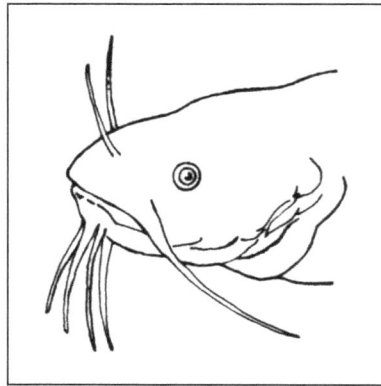
Barbels.

The head or skull includes the skull roof (a set of bones covering the brain, eyes and nostrils), the snout (from the eye to the forward most point of the upper jaw), the operculum or gill cover (absent in sharks and jawless fish), and the cheek, which extends from the eye to preopercle. The operculum and preopercle may or may not have spines. In sharks and some primitive bony fish a spiracle, small extra gill opening, is found behind each eye.

The skull in fishes is formed from a series of only loosely connected bones. Jawless fish and sharks only possess a cartilaginous endocranium, with the upper and lower jaws of cartilaginous fish being separate elements not attached to the skull. Bony fishes have additional dermal bone, forming a more or less coherent skull roof in lungfish and holost fish. The lower jaw defines a chin.

In lampreys, the mouth is formed into an oral disk. In most jawed fish, however, there are three general configurations. The mouth may be on the forward end of the head (terminal), may be upturned (superior), or may be turned downwards or on the bottom of the fish (subterminal or inferior). The mouth may be modified into a suckermouth adapted for clinging onto objects in fast-moving water.

The simpler structure is found in jawless fish, in which the cranium is represented by a trough-like basket of cartilaginous elements only partially enclosing the brain, and associated with the capsules for the inner ears and the single nostril. Distinctively, these fish have no jaws.

Cartilaginous fish, such as sharks, also have simple, and presumably primitive, skull structures. The cranium is a single structure forming a case around the brain, enclosing the lower surface and the sides, but always at least partially open at the top as a large fontanelle. The most anterior part of the cranium includes a forward plate of cartilage, the rostrum, and capsules to enclose the olfactory organs. Behind these are the orbits, and then an additional pair of capsules enclosing the structure of the inner ear. Finally, the skull tapers towards the rear, where the foramen magnum lies immediately above a single condyle, articulating with the first vertebra. There are, in addition, at various points throughout the cranium, smaller foramina for the cranial nerves. The jaws consist of separate hoops of cartilage, almost always distinct from the cranium proper.

In the ray-finned fishes, there has also been considerable modification from the primitive pattern. The roof of the skull is generally well formed, and although the exact relationship of its bones to those of tetrapods is unclear, they are usually given similar names for convenience. Other elements of the skull, however, may be reduced; there is little cheek region behind the enlarged orbits, and little, if any bone in between them. The upper jaw is often formed largely from the premaxilla, with the maxilla itself located further back, and an additional bone, the symplectic, linking the jaw to the rest of the cranium.

Although the skulls of fossil lobe-finned fish resemble those of the early tetrapods, the same cannot be said of those of the living lungfishes. The skull roof is not fully formed, and consists of multiple, somewhat irregularly shaped bones with no direct relationship to those of tetrapods. The upper jaw is formed from the pterygoids and vomers alone, all of which bear teeth. Much of the skull is formed from cartilage, and its overall structure is reduced.

The head may have several fleshy structures known as barbels, which may be very long and resemble whiskers. Many fish species also have a variety of protrusions or spines on the head. The nostrils or nares of almost all fishes do not connect to the oral cavity, but are pits of varying shape and depth.

Skull of a northern pike.

Skull of *Tiktaalik*, a genus of extinct sarcopterygian (lobe-finned "fish") from the late Devonian period.

External Organs

Jaw

The vertebrate jaw probably originally evolved in the Silurian period and appeared in the Placoderm fish which further diversified in the Devonian. Jaws are thought to derive from the pharyngeal arches that support the gills in fish. The two most anterior of these arches are thought to have become the jaw itself and the hyoid arch, which braces the jaw against the braincase and increases mechanical efficiency. While there is no fossil evidence directly to support this theory, it makes sense in light of the numbers of pharyngeal arches that are visible in extant jawed (the Gnathostomes), which have seven arches, and primitive jawless vertebrates (the Agnatha), which have nine.

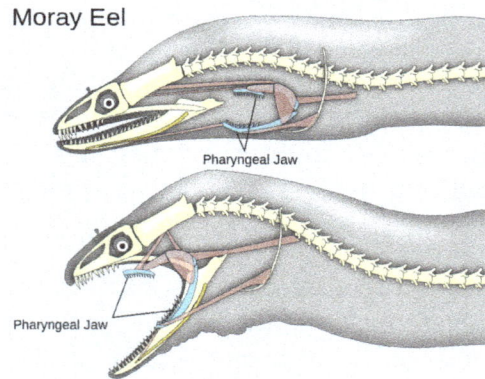

Moray eels have two sets of jaws: the oral jaws that capture prey and the pharyngeal jaws that advance into the mouth and move prey from the oral jaws to the esophagus for swallowing.

It is thought that the original selective advantage garnered by the jaw was not related to feeding, but to increased respiration efficiency. The jaws were used in the buccal pump (observable in modern fish and amphibians) that pumps water across the gills of fish or air into the lungs in the case of amphibians. Over evolutionary time the more familiar use of jaws (to humans), in feeding, was selected for and became a very important function in vertebrates.

Jaws of great white shark.

Linkage systems are widely distributed in animals. The most thorough overview of the different types of linkages in animals has been provided by M. Muller, who also designed a new classification system, which is especially well suited for biological systems. Linkage mechanisms are especially frequent and manifold in the head of bony fishes, such as wrasses, which have evolved many specialized feeding mechanisms. Especially advanced are the linkage mechanisms of jaw protrusion. For suction feeding a system of linked four-bar linkages is responsible for the coordinated opening of the mouth and 3-D expansion of the buccal cavity. Other linkages are responsible for protrusion of the premaxilla.

Eyes

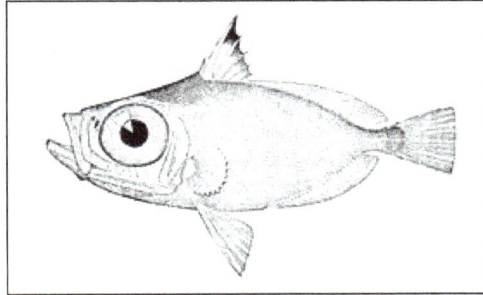

Zenion hololepis is a small deep water fish with large eyes.

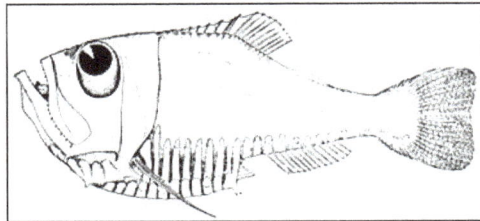

The deep sea half-naked hatchetfish has eyes which look overhead where it can see the silhouettes of prey.

Fish eyes are similar to terrestrial vertebrates like birds and mammals, but have a more spherical lens. Their retinas generally have both rod cells and cone cells (for scotopic and photopic vision), and most species have colour vision. Some fish can see ultraviolet and some can see polarized light. Amongst jawless fish, the lamprey has well-developed eyes, while the hagfish has only primitive eyespots. The ancestors of modern hagfish, thought to be the protovertebrate were evidently pushed to very deep, dark waters, where they were less vulnerable to sighted predators, and where it is advantageous to have a convex eye-spot, which gathers more light than a flat or concave one. Unlike humans, fish normally adjust focus by moving the lens closer to or further from the retina.

Gills

Gill of a rainbow trout.

The gills, located under the operculum, are a respiratory organ for the extraction of oxygen from water and for the excretion of carbon dioxide. They are not usually visible, but can be seen in some species, such as the frilled shark. The labyrinth organ of Anabantoidei and Clariidae is used to allow the fish to extract oxygen from the air. Gill rakers are bony or cartilaginous, finger-like projections off the gill arch which function in filter-feeders to retain filtered prey.

Skin

The epidermis of fish consists entirely of live cells, with only minimal quantities of keratin in the cells of the superficial layer. It is generally permeable. The dermis of bony fish typically contains relatively little of the connective tissue found in tetrapods. Instead, in most species, it is largely replaced by solid, protective bony scales. Apart from some particularly large dermal bones that form parts of the skull, these scales are lost in tetrapods, although many reptiles do have scales of a different kind, as do pangolins. Cartilaginous fish have numerous tooth-like denticles embedded in their skin, in place of true scales.

Sweat glands and sebaceous glands are both unique to mammals, but other types of skin glands are found in fish. Fish typically have numerous individual mucus-secreting skin cells that aid in insulation and protection, but may also have poison glands, photophores, or cells that produce a more watery, serous fluid. Melanin colours the skin of many species, but in fish the epidermis is often relatively colourless. Instead, the colour of the skin is largely due to chromatophores in the dermis, which, in addition to melanin, may contain guanine or carotenoid pigments. Many species, such as flounders, change the colour of their skin by adjusting the relative size of their chromatophores.

Scales

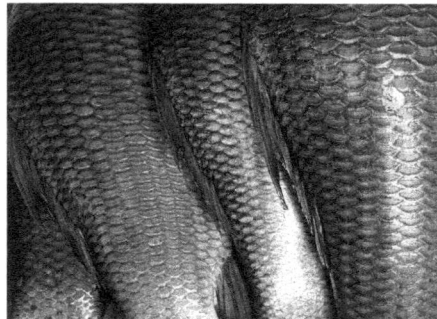

Cycloid scales covering rohu.

The outer body of many fish is covered with scales, which are part of the fish's integumentary system. The scales originate from the mesoderm (skin), and may be similar in structure to teeth. Some species are covered instead by scutes. Others have no outer covering on the skin. Most fish are covered in a protective layer of slime (mucus).

There are four principal types of fish scales:

- Placoid scales, also called dermal denticles, are similar to teeth in that they are made of dentin covered by enamel. They are typical of sharks and rays.

- Ganoid scales are flat, basal-looking scales that cover a fish body with little overlapping. They are typical of gar and bichirs.

- Cycloid scales are small oval-shaped scales with growth rings. Bowfin and remora have cycloid scales.

- Ctenoid scales are similar to the cycloid scales, with growth rings. They are distinguished by spines that cover one edge. Halibut have this type of scale.

Another, less common, type of scale is the scute, which is:

- An external shield-like bony plate, or

- A modified, thickened scale that often is keeled or spiny, or

- A projecting, modified (rough and strongly ridged) scale, usually associated with the lateral line, or on the caudal peduncle forming caudal keels, or along the ventral profile. Some fish, such as pineconefish, are completely or partially covered in scutes.

Lateral Line

The lateral line is clearly visible as a line of receptors running along the side of this Atlantic cod.

The lateral line is a sense organ used to detect movement and vibration in the surrounding water. For example, fish can use their lateral line system to follow the vortices produced by fleeing prey. In most species, it consists of a line of receptors running along each side of the fish.

Photophores

Photophores are light-emitting organs which appears as luminous spots on some fishes. The light can be produced from compounds during the digestion of prey, from specialized mitochondrial cells in the organism called photocytes, or associated with symbiotic bacteria, and are used for attracting food or confusing predators.

Fins

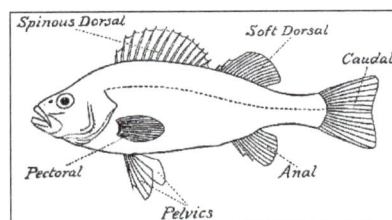

The haddock, a type of cod, is ray-finned. It has three dorsal and two anal fins.

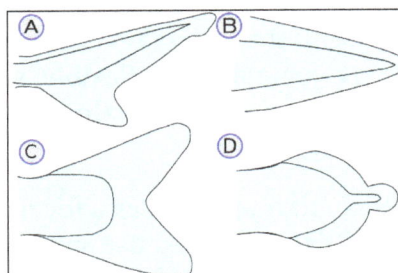

Types of caudal (tail) fin: (A) Heterocercal, (B) Protocercal, (C) Homocercal, (D) Diphycercal.

Sharks possess a heterocercal caudal fin. The dorsal portion is usually larger than the ventral portion.

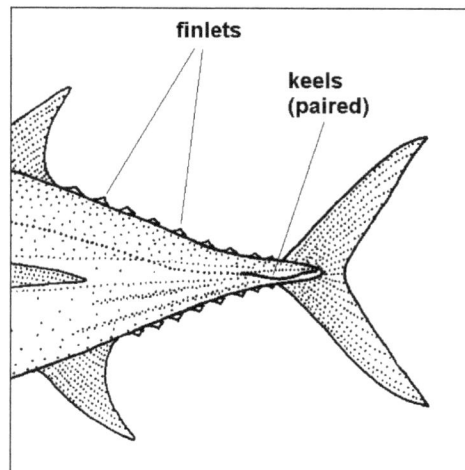

The high performance bigeye tuna is equipped with a homocercal caudal fin and finlets and keels.

Fins are the most distinctive features of fish. They are either composed of bony spines or rays protruding from the body with skin covering them and joining them together, either in a webbed fashion as seen in most bony fish or similar to a flipper as seen in sharks. Apart from the tail or caudal fin, fins have no direct connection with the spine and are supported by muscles only. Their principal function is to help the fish swim. Fins can also be used for gliding or crawling, as seen in the flying fish and frogfish. Fins located in different places on the fish serve different purposes, such as moving forward, turning, and keeping an upright position. For every fin, there are a number of fish species in which this particular fin has been lost during evolution.

Spines and Rays

In bony fish, most fins may have spines or rays. A fin may contain only spiny rays, only soft rays, or a combination of both. If both are present, the spiny rays are always anterior. Spines are generally stiff, sharp and unsegmented. Rays are generally soft, flexible, segmented, and may be branched. This segmentation of rays is the main difference that distinguishes them from spines; spines may be flexible in certain species, but never segmented.

Spines have a variety of uses. In catfish, they are used as a form of defense; many catfish have the ability to lock their spines outwards. Triggerfish also use spines to lock themselves in crevices to prevent them being pulled out.

Lepidotrichia are bony, bilaterally-paired, segmented fin rays found in bony fishes. They develop around actinotrichia as part of the dermal exoskeleton. Lepidotrichia may have some cartilage or bone in them as well. They are actually segmented and appear as a series of disks stacked one on top of another. The genetic basis for the formation of the fin rays is thought to be genes coding for the proteins actinodin 1 and actinodin 2.

Types of Fin

- Dorsal fins are located on the back. Most fishes have one dorsal fin, but some fishes have two or three. The dorsal fins serve to protect the fish against rolling, and assists in sudden turns and stops. In anglerfish, the anterior of the dorsal fin is modified into an illicium and esca, a biological equivalent to a fishing rod and lure. The bones that support the dorsal fin are called Pterygiophore. There are two to three of them: "proximal", "middle", and "distal". In spinous fins the distal is often fused to the middle, or not present at all.

- The caudal fin is the tail fin, located at the end of the caudal peduncle and is used for propulsion. The caudal peduncle is the narrow part of the fish's body to which the caudal or tail fin is attached. The hypural joint is the joint between the caudal fin and the last of the vertebrae. The hypural is often fan-shaped. The tail is called:

  - Heterocercal if the vertebrae extend into the upper lobe of the tail, making it longer (as in sharks).

  - Reversed heterocercal if the vertebrae extend into the lower lobe of the tail, making it longer (as in the Anaspida).

  - Protocercal if the vertebrae extend to the tip of the tail and the tail is symmetrical but not expanded (as in amphioxus).

  - Diphycercal if the vertebrae extend to the tip of the tail and the tail is symmetrical and expanded (as in the bichir, lungfish, lamprey and coelacanth. Most Palaeozoic fishes had a diphycercal heterocercal tail).

  - Most fish have a homocercal tail, where the fin appears superficially symmetric but the vertebrae extend for a very short distance into the upper lobe of the fin. This can be expressed in a variety of shapes. The tail fin can be:

- Rounded at the end.

- Truncated: Or end in a more-or-less vertical edge, such as in salmon.

- Forked: Or end in two prongs.

- Emarginate: Or with a slight inward curve.

- Continuous: With dorsal, caudal and anal fins attached, such as in eels.

- The anal fin is located on the ventral surface behind the anus. This fin is used to stabilize the fish while swimming.

- The paired pectoral fins are located on each side, usually just behind the operculum, and are homologous to the forelimbs of tetrapods. A peculiar function of pectoral fins, highly developed in some fish, is the creation of the dynamic lifting force that assists some fish, such as sharks, in maintaining depth and also enables the "flight" for flying fish. In many fish, the pectoral fins aid in walking, especially in the lobe-like fins of some anglerfish and in the mudskipper. Certain rays of the pectoral fins may be adapted into finger-like projections, such as in sea robins and flying gurnards. The "horns" of manta rays and their relatives are called *cephalic fins*; this is actually a modification of the anterior portion of the pectoral fin.

- The paired pelvic or ventral fins are located ventrally below the pectoral fins. They are homologous to the hindlimbs of tetrapods. The pelvic fin assists the fish in going up or down through the water, turning sharply, and stopping quickly. In gobies, the pelvic fins are often fused into a single sucker disk. This can be used to attach to objects.

- The adipose fin is a soft, fleshy fin found on the back behind the dorsal fin and just forward of the caudal fin. It is absent in many fish families, but is found in Salmonidae, characins and catfishes. Its function has remained a mystery, and is frequently clipped off to mark hatchery-raised fish, though data from 2005 showed that trout with their adipose fin removed have an 8% higher tailbeat frequency. Additional research published in 2011 has suggested that the fin may be vital for the detection of and response to stimuli such as touch, sound and changes in pressure. Canadian researchers identified a neural network in the fin, indicating that it likely has a sensory function, but are still not sure exactly what the consequences of removing it are.

- Some types of fast-swimming fish have a horizontal caudal keel just forward of the tail fin. Much like the keel of a ship, this is a lateral ridge on the caudal peduncle, usually composed of scutes, that provides stability and support to the caudal fin. There may be a single paired keel, one on each side, or two pairs above and below.

- Finlets are small fins, generally between the dorsal and the caudal fins also between the anal fin and the caudal fin (in bichirs, there are only finlets on the dorsal surface and no dorsal fin). In some fish such as tuna or sauries, they are rayless, non-retractable, and found between the last dorsal and/or anal fin and the caudal fin.

Internal organs

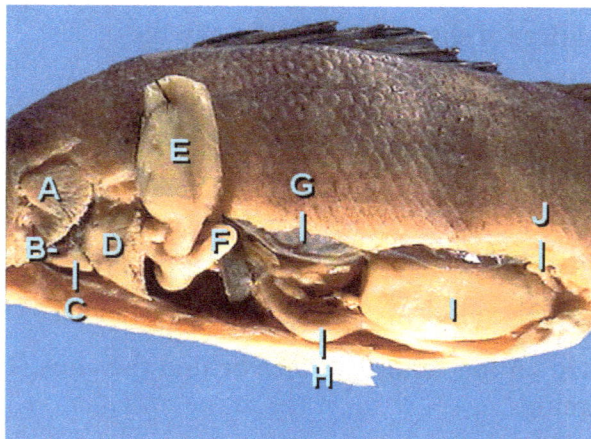

Internal organs of a male yellow perch A = gill, B = heart atrium, C: heart ventricle, D: liver (cut), E: stomach, F = pyloric caeca, G = swim bladder, H = intestine, I = testis, J = urinary bladder.

Intestines

As with other vertebrates, the intestines of fish consist of two segments, the small intestine and the large intestine. In most higher vertebrates, the small intestine is further divided into the duodenum and other parts. In fish, the divisions of the small intestine are not as clear, and the terms anterior intestine or proximal intestine may be used instead of duodenum. In bony fish, the intestine is relatively short, typically around one and a half times the length of the fish's body. It commonly has a number of pyloric caeca, small pouch-like structures along its length that help to increase the overall surface area of the organ for digesting food. There is no ileocaecal valve in teleosts, with the boundary between the small intestine and the rectum being marked only by the end of the digestive epithelium. There is no small intestine as such in non-teleost fish, such as sharks, sturgeons, and lungfish. Instead, the digestive part of the gut forms a spiral intestine, connecting the stomach to the rectum. In this type of gut, the intestine itself is relatively straight, but has a long fold running along the inner surface in a spiral fashion, sometimes for dozens of turns. This fold creates a valve-like structure that greatly increases both the surface area and the effective length of the intestine. The lining of the spiral intestine is similar to that of the small intestine in teleosts and non-mammalian tetrapods. In lampreys, the spiral valve is extremely small, possibly because their diet requires little digestion. Hagfish have no spiral valve at all, with digestion occurring for almost the entire length of the intestine, which is not subdivided into different regions.

Pyloric Caeca

The pyloric caecum is a pouch, usually peritoneal, at the beginning of the large intestine. It receives faecal material from the ileum, and connects to the ascending colon of the large intestine. It is present in most amniotes, and also in lungfish. Many fish in addition have a number of small outpocketings, also called pyloric caeca, along their intestine; despite the name they are not homologous with the caecum of amniotes. Their purpose is to increase the overall surface area of the digestive epithelium, therefore optimizing the absorption of sugars, amino acids, and dipeptides, among other nutrients.

The black swallower is a species of deep sea fish with an extensible stomach
which allows it to swallow fish larger than itself.

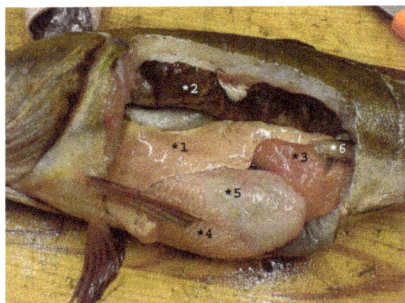

Internal organs of a female Atlantic cod: 1. Liver, 2. Gas bladder,
3. ovary, 4. Pyloric caeca, 5. Stomach, 6. Intestine

Stomach

As with other vertebrates, the relative positions of the esophageal and duodenal openings to the stomach remain relatively constant. As a result, the stomach always curves somewhat to the left before curving back to meet the pyloric sphincter. However, lampreys, hagfishes, chimaeras, lungfishes, and some teleost fish have no stomach at all, with the esophagus opening directly into the intestine. These fish consume diets that either require little storage of food, or no pre-digestion with gastric juices, or both.

Kidneys

The kidneys of fish are typically narrow, elongated organs, occupying a significant portion of the trunk. They are similar to the mesonephros of higher vertebrates (reptiles, birds and mammals). The kidneys contain clusters of nephrons, serviced by collecting ducts which usually drain into a mesonephric duct. However, the situation is not always so simple. In cartilaginous fish there is also a shorter duct which drains the posterior (metanephric) parts of the kidney, and joins with the mesonephric duct at the bladder or cloaca. Indeed, in many cartilaginous fish, the anterior portion of the kidney may degenerate or cease to function altogether in the adult. Hagfish and lamprey kidneys are unusually simple. They consist of a row of nephrons, each emptying directly into the mesonephric duct.

Spleen

The spleen is found in nearly all vertebrates. It is a non-vital organ, similar in structure to a large lymph node. It acts primarily as a blood filter, and plays important roles in regard to red blood cells and the immune system. In cartilaginous and bony fish it consists primarily of red pulp and is normally a somewhat elongated organ as it actually lies inside the serosal lining of the intestine. The only vertebrates lacking a spleen are the lampreys and hagfishes. Even in these animals, there is a diffuse layer of haematopoeitic tissue within the gut wall, which has a similar structure to red pulp, and is presumed to be homologous with the spleen of higher vertebrates.

Liver

The liver is a large vital organ present in all fish. It has a wide range of functions, including detoxification, protein synthesis, and production of biochemicals necessary for digestion. It is very susceptible to contamination by organic and inorganic compounds because they can accumulate over time and cause potentially life-threatening conditions. Because of the liver's capacity for detoxification and storage of harmful components, it is often used as an environmental biomarker.

Heart

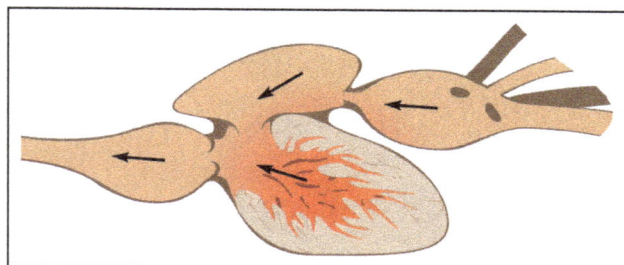

Blood flow through the heart: sinus venosus, atrium, ventricle, and outflow tract.

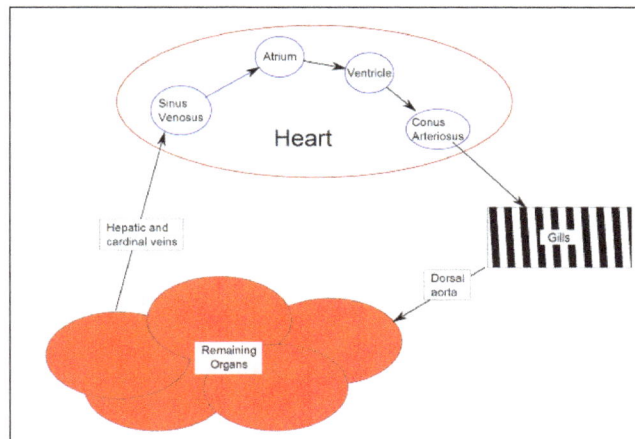

Cardiovascular cycle in a fish.

Fish have what is often described as a two-chambered heart, consisting of one atrium to receive blood and one ventricle to pump it, in contrast to three chambers (two atria, one ventricle) of amphibian and most reptile hearts and four chambers (two atria, two ventricles) of mammal and bird hearts. However, the fish heart has entry and exit compartments that may be called chambers, so it is also sometimes described as three-chambered or four-chambered, depending on what is counted as a chamber. The atrium and ventricle are sometimes considered "true chambers", while the others are considered "accessory chambers".

The four compartments are arranged sequentially:

- Sinus venosus, a thin-walled sac or reservoir with some cardiac muscle that collects deoxygenated blood through the incoming hepatic and cardinal veins.

- Atrium, a thicker-walled, muscular chamber that sends blood to the ventricle.

- Ventricle, a thick-walled, muscular chamber that pumps the blood to the fourth part, the outflow tract. The shape of the ventricle varies considerably, usually tubular in fish with elongated bodies, pyramidal with a triangular base in others, or sometimes sac-like in some marine fish.

- The outflow tract (OFT) to the ventral aorta, consisting of the tubular conus arteriosus, bulbus arteriosus, or both. The conus arteriosus, typically found in more primitive species of fish, contracts to assist blood flow to the aorta, while the bulbus anteriosus does not.

Ostial valves, consisting of flap-like connective tissues, prevent blood from flowing backward through the compartments. The ostial valve between the sinus venosus and atrium is called the sino-atrial valve, which closes during ventricular contraction. Between the atrium and ventricle is an ostial valve called the atrio-ventricular valve, and between the bulbus arteriosus and ventricle is an ostial valve called the bulbo-ventricular valve. The conus arteriosus has a variable number of semilunar valves.

The ventral aorta delivers blood to the gills where it is oxygenated and flows, through the dorsal aorta, into the rest of the body. (In tetrapods, the ventral aorta has divided in two; one half forms the ascending aorta, while the other forms the pulmonary artery).

The circulatory systems of all vertebrates, are *closed*. Fish have the simplest circulatory system, consisting of only one circuit, with the blood being pumped through the capillaries of the gills and on to the capillaries of the body tissues. This is known as *single cycle* circulation.

In the adult fish, the four compartments are not arranged in a straight row but, instead form an S-shape with the latter two compartments lying above the former two. This relatively simpler pattern is found in cartilaginous fish and in the ray-finned fish. In teleosts, the conus arteriosus is very small and can more accurately be described as part of the aorta rather than of the heart proper. The conus arteriosus is not present in any amniotes, presumably having been absorbed into the ventricles over the course of evolution. Similarly, while the sinus venosus is present as a vestigial structure in some reptiles and birds, it is otherwise absorbed into the right atrium and is no longer distinguishable.

Swim Bladder

The swim bladder of a rudd.

The swim bladder (or gas bladder) is an internal organ that contributes to the ability of a fish to control its buoyancy, and thus to stay at the current water depth, ascend, or descend without having to waste energy in swimming. The bladder is found only in the bony fishes. In the more primitive groups like some minnows, bichirs and lungfish, the bladder is open to the esophagus and doubles as a lung. It is often absent in fast swimming fishes such as the tuna and mackerel families. The condition of a bladder open to the esophagus is called physostome, the closed condition physoclist. In the latter, the gas content of the bladder is controlled through a rete mirabilis, a network of blood vessels effecting gas exchange between the bladder and the blood.

Weberian Apparatus

Fishes of the superorder Ostariophysi possess a structure called the Weberian apparatus, a modification which allow them to hear better. This ability which may well explain the marked success of otophysian fishes. The apparatus is made up of a set of bones known as *Weberian ossicles*, a chain of small bones that connect the auditory system to the swim bladder of fishes. The ossicles connect the gas bladder wall with Y-shaped lymph sinus that abuts the lymph-filled transverse canal joining the sacculi of the right and left ears. This allows the transmission of vibrations to the inner ear. A fully functioning Weberian apparatus consists of the swim bladder, the Weberian ossicles, a portion of the anterior vertebral column, and some muscles and ligaments.

Reproductive Organs

Fish reproductive organs include testes and ovaries. In most species, gonads are paired organs of similar size, which can be partially or totally fused. There may also be a range of secondary organs that increase reproductive fitness. The genital papilla is a small, fleshy tube behind the anus in some fishes, from which the sperm or eggs are released; the sex of a fish often can be determined by the shape of its papilla.

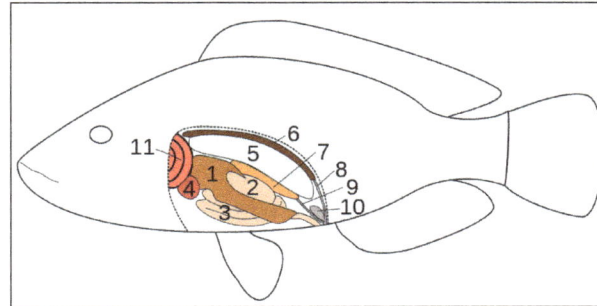

7 = testicles or ovaries.

Testes

Most male fish have two testes of similar size. In the case of sharks, the testis on the right side is usually larger. The primitive jawless fish have only a single testis, located in the midline of the body, although even this forms from the fusion of paired structures in the embryo.

Under a tough membranous shell, the tunica albuginea, the testis of some teleost fish, contains very fine coiled tubes called seminiferous tubules. The tubules are lined with a layer of cells (germ cells) that from puberty into old age, develop into sperm cells (also known as spermatozoa or male gametes). The developing sperm travel through the seminiferous tubules to the rete testis located in the mediastinum testis, to the efferent ducts, and then to the epididymis where newly created sperm cells mature. The sperm move into the vas deferens, and are eventually expelled through the urethra and out of the urethral orifice through muscular contractions.

However, most fish do not possess seminiferous tubules. Instead, the sperm are produced in spherical structures called *sperm ampullae*. These are seasonal structures, releasing their contents during the breeding season, and then being reabsorbed by the body. Before the next breeding season, new sperm ampullae begin to form and ripen. The ampullae are otherwise essentially identical to the seminiferous tubules in higher vertebrates, including the same range of cell types.

In terms of spermatogonia distribution, the structure of teleosts testes has two types: in the most common, spermatogonia occur all along the seminiferous tubules, while in Atherinomorph fish they are confined to the distal portion of these structures. Fish can present cystic or semi-cystic spermatogenesis in relation to the release phase of germ cells in cysts to the seminiferous tubules lumen.

Ovaries

Many of the features found in ovaries are common to all vertebrates, including the presence of follicular cells and tunica albuginea There may be hundreds or even millions of fertile eggs present in the ovary of a fish at any given time. Fresh eggs may be developing from the germinal epithelium throughout life. Corpora lutea are found only in mammals, and in some elasmobranch fish; in other species, the remnants of the follicle are quickly resorbed by the ovary. The ovary of teleosts is often contains a hollow, lymph-filled space which opens into the oviduct, and into which the eggs are shed. Most normal female fish have two ovaries. In some elasmobranchs, only the right ovary develops fully. In the primitive jawless fish, and some teleosts, there is only one ovary, formed by the fusion of the paired organs in the embryo.

Fish ovaries may be of three types: gymnovarian, secondary gymnovarian or cystovarian. In the first type, the oocytes are released directly into the coelomic cavity and then enter the ostium, then through the oviduct and are eliminated. Secondary gymnovarian ovaries shed ova into the coelom from which they go directly into the oviduct. In the third type, the oocytes are conveyed to the exterior through the oviduct. Gymnovaries are the primitive condition found in lungfish, sturgeon, and bowfin. Cystovaries characterize most teleosts, where the ovary lumen has continuity with the oviduct. Secondary gymnovaries are found in salmonids and a few other teleosts.

Nervous System

Central Nervous System

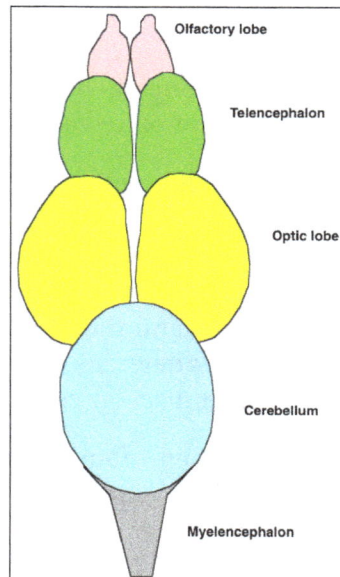

Dorsal view of the brain of the rainbow trout.

Fish typically have quite small brains relative to body size compared with other vertebrates, typically one-fifteenth the brain mass of a similarly sized bird or mammal. However, some fish have relatively large brains, most notably mormyrids and sharks, which have brains about as massive relative to body weight as birds and marsupials.

Fish brains are divided into several regions. At the front are the olfactory lobes, a pair of structures that receive and process signals from the nostrils via the two olfactory nerves. Similar to the way humans smell chemicals in the air, fish smell chemicals in the water by tasting them. The olfactory lobes are very large in fish that hunt primarily by smell, such as hagfish, sharks, and catfish. Behind the olfactory lobes is the two-lobed telencephalon, the structural equivalent to the cerebrum in higher vertebrates. In fish the telencephalon is concerned mostly with olfaction. Together these structures form the forebrain.

The forebrain is connected to the midbrain via the diencephalon (in the diagram, this structure is below the optic lobes and consequently not visible). The diencephalon performs functions associated with hormones and homeostasis. The pineal body lies just above the diencephalon. This structure detects light, maintains circadian rhythms, and controls color changes. The midbrain or mesencephalon contains the two optic lobes. These are very large in species that hunt by sight, such as rainbow trout and cichlids.

The hindbrain or metencephalon is particularly involved in swimming and balance. The cerebellum is a single-lobed structure that is typically the biggest part of the brain. Hagfish and lampreys have relatively small cerebellae, while the mormyrid cerebellum is massive and apparently involved in their electrical sense.

The brain stem or myelencephalon is the brain's posterior. As well as controlling some muscles and body organs, in bony fish at least, the brain stem governs respiration and osmoregulation.

Vertebrates are the only chordate group to exhibit a proper brain. A slight swelling of the anterior end of the dorsal nerve cord is found in the lancelet, though it lacks the eyes and other complex sense organs comparable to those of vertebrates. Other chordates do not show any trends towards cephalisation. The central nervous system is based on a hollow nerve tube running along the length of the animal, from which the peripheral nervous system branches out to innervate the various systems. The front end of the nerve tube is expanded by a thickening of the walls and expansion of the central canal of spinal cord into three primary brain vesicles: The prosencephalon (forebrain), mesencephalon (midbrain) and rhombencephalon (hindbrain), further differentiated in the various vertebrate groups. Two laterally placed eyes form around outgrows from the midbrain, except in hagfish, though this may be a secondary loss. The forebrain is well developed and subdivided in most tetrapods, while the midbrain dominate in many fish and some salamanders. Vesicles of the forebrain are usually paired, giving rise to hemispheres like the cerebral hemispheres in mammals. The resulting anatomy of the central nervous system, with a single, hollow ventral nerve cord topped by a series of (often paired) vesicles is unique to vertebrates.

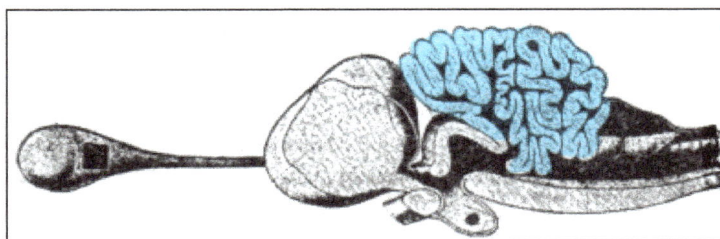

Cross-section of the brain of a porbeagle shark, with the cerebellum.

Cerebellum

The circuits in the cerebellum are similar across all classes of vertebrates, including fish, reptiles, birds, and mammals. There is also an analogous brain structure in cephalopods with well-developed brains, such as octopuses. This has been taken as evidence that the cerebellum performs functions important to all animal species with a brain.

There is considerable variation in the size and shape of the cerebellum in different vertebrate species. In amphibians, lampreys, and hagfish, the cerebellum is little developed; in the latter two groups, it is barely distinguishable from the brain-stem. Although the spinocerebellum is present in these groups, the primary structures are small paired nuclei corresponding to the vestibulocerebellum.

The cerebellum of cartilaginous and bony fishes is extraordinarily large and complex. In at least one important respect, it differs in internal structure from the mammalian cerebellum: The fish cerebellum does not contain discrete deep cerebellar nuclei. Instead, the primary targets of Purkinje cells are a distinct type of cell distributed across the cerebellar cortex, a type not seen in mammals. In mormyrid fish (a family of weakly electrosensitive freshwater fish), the cerebellum is considerably larger than the rest of the brain put together. The largest part of it is a special structure called the *valvula*, which has an unusually regular architecture and receives much of its input from the electrosensory system.

Most species of fish and amphibians possess a lateral line system that senses pressure waves in water. One of the brain areas that receives primary input from the lateral line organ, the medial octavolateral nucleus, has a cerebellum-like structure, with granule cells and parallel fibers. In electrosensitive fish, the input from the electrosensory system goes to the dorsal octavolateral nucleus, which also has a cerebellum-like structure. In ray-finned fishes (by far the largest group), the optic tectum has a layer — the marginal layer — that is cerebellum-like.

Identified Neurons

A neuron is called *identified* if it has properties that distinguish it from every other neuron in the same animal—properties such as location, neurotransmitter, gene expression pattern, and connectivity—and if every individual organism belonging to the same species has one and only one neuron with the same set of properties. In vertebrate nervous systems very few neurons are "identified" in this sense—in humans, there are believed to be none—but in simpler nervous systems, some or all neurons may be thus unique.

In vertebrates, the best known identified neurons are the gigantic Mauthner cells of fish. Every fish has two Mauthner cells, located in the bottom part of the brainstem, one on the left side and one on the right. Each Mauthner cell has an axon that crosses over, innervating neurons at the same brain level and then travelling down through the spinal cord, making numerous connections as it goes. The synapses generated by a Mauthner cell are so powerful that a single action potential gives rise to a major behavioral response: within milliseconds the fish curves its body into a C-shape, then straightens, thereby propelling itself rapidly forward. Functionally this is a fast escape response, triggered most easily by a strong sound wave or pressure wave impinging on the lateral line organ of the fish. Mauthner cells are not the only identified neurons in fish—there are about 20 more types, including pairs of "Mauthner cell analogs" in each spinal segmental nucleus. Although a Mauthner cell is capable of bringing about an escape response all by itself, in the context of ordinary behavior other types of cells usually contribute to shaping the amplitude and direction of the response.

Mauthner cells have been described as command neurons. A command neuron is a special type of identified neuron, defined as a neuron that is capable of driving a specific behavior all by itself. Such neurons appear most commonly in the fast escape systems of various species—the squid giant axon and squid giant synapse, used for pioneering experiments in neurophysiology because of their enormous size, both participate in the fast escape circuit of the squid. The concept of a command neuron has, however, become controversial, because of studies showing that some neurons that initially appeared to fit the description were really only capable of evoking a response in a limited set of circumstances.

Immune System

Immune organs vary by type of fish. In the jawless fish (lampreys and hagfish), true lymphoid organs are absent. These fish rely on regions of lymphoid tissue within other organs to produce immune cells. For example, erythrocytes, macrophages and plasma cells are produced in the anterior kidney (or pronephros) and some areas of the gut (where granulocytes mature.) They resemble primitive bone marrow in hagfish. Cartilaginous fish (sharks and rays) have a more advanced immune system. They have three specialized organs that are unique to chondrichthyes; the epigonal organs (lymphoid tissue similar to mammalian bone) that surround the gonads, the Leydig's organ within the walls of their esophagus, and a spiral valve in their intestine. These organs house typical immune cells (granulocytes, lymphocytes and plasma cells). They also possess an identifiable thymus and a well-developed spleen (their most important immune organ) where various lymphocytes, plasma cells and macrophages develop and are stored. Chondrostean fish (sturgeons, paddlefish and bichirs) possess a major site for the production of granulocytes within a mass that is associated with the meninges (membranes surrounding the central nervous system.) Their heart is frequently covered with tissue that contains lymphocytes, reticular cells and a small number of macrophages. The chondrostean kidney is an important hemopoietic organ; where erythrocytes, granulocytes, lymphocytes and macrophages develop.

Like chondrostean fish, the major immune tissues of bony fish (or teleostei) include the kidney (especially the anterior kidney), which houses many different immune cells. In addition, teleost fish possess a thymus, spleen and scattered immune areas within mucosal tissues (e.g. in the skin, gills, gut and gonads). Much like the mammalian immune system, teleost erythrocytes, neutrophils and granulocytes are believed to reside in the spleen whereas lymphocytes are the major cell type found in the thymus. In 2006, a lymphatic system similar to that in mammals was described in one species of teleost fish, the zebrafish. Although not confirmed as yet, this system presumably will be where naive (unstimulated) T cells accumulate while waiting to encounter an antigen.

Fish Physiology

Fish physiology is the scientific study of how the component parts of fish function together in the living fish. It can be contrasted with fish anatomy, which is the study of the form or morphology of fishes. In practice, fish anatomy and physiology complement each other, the former dealing with the structure of a fish, its organs or component parts and how they are put together, such as might be observed on the dissecting table or under the microscope, and the later dealing with how those components function together in the living fish.

Circulation

The circulatory systems of all vertebrates are *closed*, just as in humans. Still, the systems of fish, amphibians, reptiles, and birds show various stages of the evolution of the circulatory system. In fish, the system has only one circuit, with the blood being pumped through the capillaries of the gills and on to the capillaries of the body tissues. This is known as *single cycle* circulation. The heart of fish is therefore only a single pump (consisting of two chambers). Fish have a closed-loop circulatory system. The heart pumps the blood in a single loop throughout the body. In most fish, the heart consists of four parts, including two chambers and an entrance and exit. The first part is the sinus venosus, a thin-walled sac that collects blood from the fish's veins before allowing it to flow to the second part, the atrium, which is a large muscular chamber. The atrium serves as a one-way antechamber, sends blood to the third part, ventricle. The ventricle is another thick-walled, muscular chamber and it pumps the blood, first to the fourth part, bulbus arteriosus, a large tube, and then out of the heart. The bulbus arteriosus connects to the aorta, through which blood flows to the gills for oxygenation.

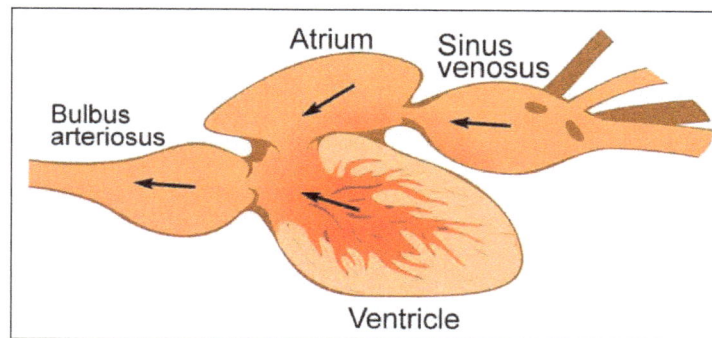

Two-chambered heart of a fish.

In amphibians and most reptiles, a double circulatory system is used, but the heart is not always completely separated into two pumps. Amphibians have a three-chambered heart.

Digestion

Jaws allow fish to eat a wide variety of food, including plants and other organisms. Fish ingest food through the mouth and break it down in the esophagus. In the stomach, food is further digested and, in many fish, processed in finger-shaped pouches called pyloric caeca, which secrete digestive enzymes and absorb nutrients. Organs such as the liver and pancreas add enzymes and various chemicals as the food moves through the digestive tract. The intestine completes the process of digestion and nutrient absorption.

In most vertebrates, digestion is a four-stage process involving the main structures of the digestive tract, starting with ingestion, placing food into the mouth, and concluding with the excretion of undigested material through the anus. From the mouth, the food moves to the stomach, where as bolus it is broken down chemically. It then moves to the intestine, where the process of breaking the food down into simple molecules continues and the results are absorbed as nutrients into the circulatory and lymphatic system.

Although the precise shape and size of the stomach varies widely among different vertebrates, the relative positions of the oesophageal and duodenal openings remain relatively constant. As a result, the organ always curves somewhat to the left before curving back to meet the pyloric sphincter. However, lampreys, hagfishes, chimaeras, lungfishes, and some teleost fish have no stomach at all, with the oesophagus opening directly into the intestine. These animals all consume diets that either require little storage of food, or no pre-digestion with gastric juices, or both.

The small intestine is the part of the digestive tract following the stomach and followed by the large intestine, and is where much of the digestion and absorption of food takes place. In fish, the divisions of the small intestine are not clear, and the terms *anterior* or *proximal* intestine may be used instead of duodenum. The small intestine is found in all teleosts, although its form and length vary enormously between species. In teleosts, it is relatively short, typically around one and a half times the length of the fish's body. It commonly has a number of *pyloric caeca*, small pouch-like structures along its length that help to increase the overall surface area of the organ for digesting food. There is no ileocaecal valve in teleosts, with the boundary between the small intestine and the rectum being marked only by the end of the digestive epithelium.

There is no small intestine as such in non-teleost fish, such as sharks, sturgeons, and lungfish. Instead, the digestive part of the gut forms a spiral intestine, connecting the stomach to the rectum. In this type of gut, the intestine itself is relatively straight, but has a long fold running along the inner surface in a spiral fashion, sometimes for dozens of turns. This valve greatly increases both the surface area and the effective length of the intestine. The lining of the spiral intestine is similar to that of the small intestine in teleosts and non-mammalian tetrapods. In lampreys, the spiral valve is extremely small, possibly because their diet requires little digestion. Hagfish have no spiral valve at all, with digestion occurring for almost the entire length of the intestine, which is not subdivided into different regions.

The large intestine is the last part of the digestive system normally found in vertebrate animals. Its function is to absorb water from the remaining indigestible food matter, and then to pass useless waste material from the body. In fish, there is no true large intestine, but simply a short rectum connecting the end of the digestive part of the gut to the cloaca. In sharks, this includes a *rectal gland* that secretes salt to help the animal maintain osmotic balance with the seawater. The gland somewhat resembles a caecum in structure, but is not a homologous structure.

As with many aquatic animals, most fish release their nitrogenous wastes as ammonia. Some of the wastes diffuse through the gills. Blood wastes are filtered by the kidneys.

Saltwater fish tend to lose water because of osmosis. Their kidneys return water to the body. The reverse happens in freshwater fish: they tend to gain water osmotically. Their kidneys produce dilute urine for excretion. Some fish have specially adapted kidneys that vary in function, allowing them to move from freshwater to saltwater.

In sharks, digestion can take a long time. The food moves from the mouth to a J-shaped stomach, where it is stored and initial digestion occurs. Unwanted items may never get past the stomach, and instead the shark either vomits or turns its stomachs inside out and ejects unwanted items from its mouth. One of the biggest differences between the digestive systems of sharks and mammals is that sharks have much shorter intestines. This short length is achieved by the spiral valve with multiple turns within a single short section instead of a long tube-like intestine. The valve provides a long surface area, requiring food to circulate inside the short gut until fully digested, when remaining waste products pass into the cloaca.

Endocrine System

Regulation of Social Behavior

Oxytocin is a group of neuropeptides found in most vertebrate. One form of oxytocin functions as a hormone which is associated with human love. In 2012, researchers injected cichlids from the social species *Neolamprologus pulcher*, either with this form of isotocin or with a control saline solution. They found isotocin increased "responsiveness to social information", which suggests "it is a key regulator of social behavior that has evolved and endured since ancient times".

Effects of Pollution

Fish can bioaccumulate pollutants that are discharged into waterways. Estrogenic compounds found in pesticides, birth control, plastics, plants, fungi, bacteria, and synthetic drugs leeched into rivers are affecting the endocrine systems of native species. In Boulder, Colorado, white sucker fish found downstream of a municipal waste water treatment plant exhibit impaired or abnormal sexual development. The fish have been exposed to higher levels of estrogen, and leading to feminized fish. Males display female reproductive organs, and both sexes have reduced fertility, and a higher hatch mortality.

Freshwater habitats in the United States are widely contaminated by the common pesticide atrazine. There is controversy over the degree to which this pesticide harms the endocrine systems of freshwater fish and amphibians. Non-industry-funded researchers consistently report harmful effects while industry-funded researchers consistently report no harmful effects.

In the marine ecosystem, organochlorine contaminants like pesticides, herbicides (DDT), and chlordan are accumulating within fish tissue and disrupting their endocrine system. High frequencies of infertility and high levels of organochlorines have been found in bonnethead sharks along the Gulf Coast of Florida. These endocrine-disrupting compounds are similar in structure to naturally occurring hormones in fish. They can modulate hormonal interactions in fish by:

- Binding to cellular receptors, causing unpredictable and abnormal cell activity.

- Blocking receptor sites, inhibiting activity.

- Promoting the creation of extra receptor sites, amplifying the effects of the hormone or compound.

- Interacting with naturally occurring hormones, changing their shape and impact

- Affecting hormone synthesis or metabolism, causing an improper balance or quantity of hormones.

Osmoregulation

Two major types of osmoregulation are osmoconformers and osmoregulators. Osmoconformers match their body osmolarity to their environment actively or passively. Most marine invertebrates are osmoconformers, although their ionic composition may be different from that of seawater.

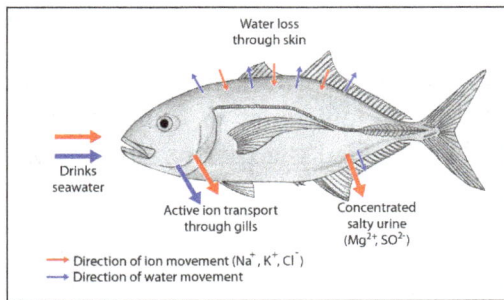

Movement of water and ions in saltwater fish.

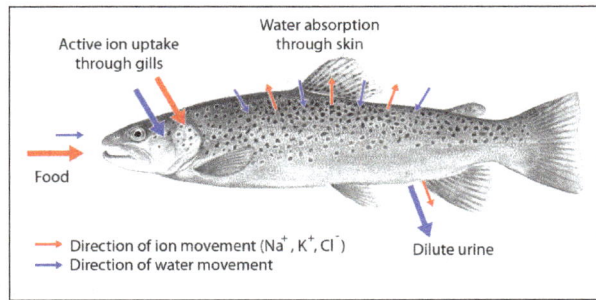

Movement of water and ions in freshwater fish.

Osmoregulators tightly regulate their body osmolarity, which always stays constant, and are more common in the animal kingdom. Osmoregulators actively control salt concentrations despite the salt concentrations in the environment. An example is freshwater fish. The gills actively uptake salt from the environment by the use of mitochondria-rich cells. Water will diffuse into the fish, so it excretes a very hypotonic (dilute) urine to expel all the excess water. A marine fish has an internal osmotic concentration lower than that of the surrounding seawater, so it tends to lose water and gain salt. It actively excretes salt out from the gills. Most fish are stenohaline, which means they are restricted to either salt or fresh water and cannot survive in water with a different salt concentration than they are adapted to. However, some fish show a tremendous ability to effectively osmoregulate across a broad range of salinities; fish with this ability are known as euryhaline species, e.g., salmon. Salmon has been observed to inhabit two utterly disparate environments — marine and fresh water — and it is inherent to adapt to both by bringing in behavioral and physiological modifications.

In contrast to bony fish, with the exception of the coelacanth, the blood and other tissue of sharks and Chondrichthyes is generally isotonic to their marine environments because of the high concentration of urea and trimethylamine N-oxide (TMAO), allowing them to be in osmotic balance with the seawater. This adaptation prevents most sharks from surviving in freshwater, and they are therefore confined to marine environments. A few exceptions exist, such as the bull shark, which has developed a way to change its kidney function to excrete large amounts of urea. When a shark dies, the urea is broken down to ammonia by bacteria, causing the dead body to gradually smell strongly of ammonia.

Sharks have adopted a different, efficient mechanism to conserve water, i.e., osmoregulation. They retain urea in their blood in relatively higher concentration. Urea is damaging to living tissue so, to cope with this problem, some fish retain *trimethylamine oxide*. This provides a better solution to urea's toxicity. Sharks, having slightly higher solute concentration (i.e., above 1000 mOsm which is sea solute concentration), do not drink water like fresh water fish.

Thermoregulation

Homeothermy and poikilothermy refer to how stable an organism's temperature is. Most endothermic organisms are homeothermic, like mammals. However, animals with facultative endothermy are often poikilothermic, meaning their temperature can vary considerably. Similarly, most fish are ectotherms, as all of their heat comes from the surrounding water. However, most are homeotherms because their temperature is very stable.

Most organisms have a preferred temperature range, however some can be acclimated to temperatures colder or warmer than what they are typically used to. An organism's preferred temperature is typically the temperature at which the organism's physiological processes can act at optimal rates. When fish become acclimated to other temperatures, the efficiency of their physiological processes may decrease but will continue to function. This is called the thermal neutral zone at which an organism can survive indefinitely.

H.M. Vernon has done work on the death temperature and paralysis temperature (temperature of heat rigor) of various animals. He found that species of the same class showed very similar temperature values, those from the Amphibia examined being 38.5 °C, fish 39 °C, Reptilia 45 °C, and various Molluscs 46 °C.

To cope with low temperatures, some fish have developed the ability to remain functional even when the water temperature is below freezing; some use natural antifreeze or antifreeze proteins to resist ice crystal formation in their tissues.

Most sharks are "cold-blooded" or, more precisely, poikilothermic, meaning that their internal body temperature matches that of their ambient environment. Members of the family Lamnidae (such as the shortfin mako shark and the great white shark) are homeothermic and maintain a higher body temperature than the surrounding water. In these sharks, a strip of aerobic red muscle located near the center of the body generates the heat, which the body retains via a countercurrent exchange mechanism by a system of blood vessels called the rete mirabile ("miraculous net"). The common thresher shark has a similar mechanism for maintaining an elevated body temperature, which is thought to have evolved independently.

Tuna can maintain the temperature of certain parts of their body above the temperature of ambient seawater. For example, bluefin tuna maintain a core body temperature of 25–33 °C (77–91 °F), in water as cold as 6 °C (43 °F). However, unlike typical endothermic creatures such as mammals and birds, tuna do not maintain temperature within a relatively narrow range. Tuna achieve endothermy by conserving the heat generated through normal metabolism. The rete mirabile ("wonderful net"), the intertwining of veins and arteries in the body's periphery, transfers heat from venous blood to arterial blood via a counter-current exchange system, thus mitigating the effects of surface cooling. This allows the tuna to elevate the temperatures of the highly aerobic tissues of the skeletal muscles, eyes and brain, which supports faster swimming speeds and reduced energy expenditure, and which enables them to survive in cooler waters over a wider range of ocean environments than those of other fish. In all tunas, however, the heart operates at ambient temperature, as it receives cooled blood, and coronary circulation is directly from the gills.

- Homeothermy: Although most fish are exclusively ectothermic, there are exceptions. Certain species of fish maintain elevated body temperatures. Endothermic teleosts (bony fish) are all in the suborder Scombroidei and include the billfishes, tunas, including a "primitive" mackerel species, *Gasterochisma melampus*. All sharks in the family Lamnidae – shortfin mako, long fin mako, white, porbeagle, and salmon shark – are endothermic, and evidence suggests the trait exists in family Alopiidae (thresher sharks). The degree of endothermy varies from the billfish, which warm only their eyes and brain, to bluefin tuna and porbeagle sharks who maintain body temperatures elevated in excess of 20 °C above ambient water temperatures. Endothermy, though metabolically costly, is thought to provide advantages such as increased muscle strength, higher rates of central nervous system processing, and higher rates of digestion.

In some fish, a rete mirabile allows for an increase in muscle temperature in regions where this network of vein and arteries is found. The fish is able to thermoregulate certain areas of their body. Additionally, this increase in temperature leads to an increase in basal metabolic temperature. The fish is now able to split ATP at a higher rate and ultimately can swim faster.

The eye of a swordfish can generate heat to better cope with detecting their prey at depths of 2000 feet.

Buoyancy

Sharks, like this three tonne great white shark, don't have swim bladders.
Most sharks need to keep swimming to avoid sinking.

The body of a fish is denser than water, so fish must compensate for the difference or they will sink. Many bony fishes have an internal organ called a swim bladder, or gas bladder, that adjusts their buoyancy through manipulation of gases. In this way, fish can stay at the current water depth, or ascend or descend without having to waste energy in swimming. The bladder is only found in bony fishes. In the more primitive groups like some minnows, bichirs and lungfish, the bladder is open to the esophagus and double as a lung. It is often absent in fast swimming fishes such as the tuna and mackerel families. The condition of a bladder open to the esophagus is called physostome, the closed condition physoclist. In the latter, the gas content of the bladder is controlled through the rete mirabilis, a network of blood vessels effecting gas exchange between the bladder and the blood.

In some fish, a rete mirabile fills the swim bladder with oxygen. A countercurrent exchange system is utilized between the venous and arterial capillaries. By lowering the pH levels in the venous capillaries, oxygen unbinds from blood hemoglobin. This causes an increase in venous blood oxygen concentration, allowing the oxygen to diffuse through the capillary membrane and into the arterial capillaries, where oxygen is still sequestered to hemoglobin. The cycle of diffusion continues until the concentration of oxygen in the arterial capillaries is supersaturated (larger than the concentration of oxygen in the swim bladder). At this point, the free oxygen in the arterial capillaries diffuses into the swim bladder via the gas gland.

Unlike bony fish, sharks do not have gas-filled swim bladders for buoyancy. Instead, sharks rely on a large liver filled with oil that contains squalene, and their cartilage, which is about half the normal density of bone. Their liver constitutes up to 30% of their total body mass. The liver's effectiveness is limited, so sharks employ dynamic lift to maintain depth when not swimming. Sand tiger sharks store air in their stomachs, using it as a form of swim bladder. Most sharks need to constantly swim in order to breathe and cannot sleep very long without sinking (if at all). However, certain species, like the nurse shark, are capable of pumping water across their gills, allowing them to rest on the ocean bottom.

Reproductive Processes

Oogonia development in teleosts fish varies according to the group, and the determination of oogenesis dynamics allows the understanding of maturation and fertilisation processes. Changes in the nucleus, ooplasm, and the surrounding layers characterize the oocyte maturation process.

Postovulatory follicles are structures formed after oocyte release; they do not have endocrine function, present a wide irregular lumen, and are rapidly reabsorbed in a process involving the apoptosis of follicular cells. A degenerative process called follicular atresia reabsorbs vitellogenic oocytes not spawned. This process can also occur, but less frequently, in oocytes in other development stages.

Some fish are hermaphrodites, having both testes and ovaries either at different phases in their life cycle or, as in hamlets, have them simultaneously.

Over 97% of all known fish are oviparous, that is, the eggs develop outside the mother's body. Examples of oviparous fish include salmon, goldfish, cichlids, tuna, and eels. In the majority of these species, fertilisation takes place outside the mother's body, with the male and female fish shedding their gametes into the surrounding water. However, a few oviparous fish practice internal fertilisation, with the male using some sort of intromittent organ to deliver sperm into the genital opening of the female, most notably the oviparous sharks, such as the horn shark, and oviparous rays, such as skates. In these cases, the male is equipped with a pair of modified pelvic fins known as claspers.

Marine fish can produce high numbers of eggs which are often released into the open water column. The eggs have an average diameter of 1 millimetre (0.039 in). The eggs are generally surrounded by the extraembryonic membranes but do not develop a shell, hard or soft, around these membranes. Some fish have thick, leathery coats, especially if they must withstand physical force or desiccation. These type of eggs can also be very small and fragile.

The newly hatched young of oviparous fish are called larvae. They are usually poorly formed, carry a large yolk sac (for nourishment) and are very different in appearance from juvenile and adult specimens. The larval period in oviparous fish is relatively short (usually only several weeks), and larvae rapidly grow and change appearance and structure (a process termed metamorphosis) to become juveniles. During this transition larvae must switch from their yolk sac to feeding on zooplankton prey, a process which depends on typically inadequate zooplankton density, starving many larvae.

Egg of lamprey.   Egg of catshark   Egg of bullhead shark.   Egg of chimaera.
(mermaids' purse).

In ovoviviparous fish the eggs develop inside the mother's body after internal fertilisation but receive little or no nourishment directly from the mother, depending instead on the yolk. Each embryo develops in its own egg. Familiar examples of ovoviviparous fish include guppies, angel sharks, and coelacanths.

Some species of fish are viviparous. In such species the mother retains the eggs and nourishes the embryos. Typically, viviparous fish have a structure analogous to the placenta seen in mammals connecting the mother's blood supply with that of the embryo. Examples of viviparous fish include the surf-perches, splitfins, and lemon shark. Some viviparous fish exhibit oophagy, in which the developing embryos eat other eggs produced by the mother. This has been observed primarily among sharks, such as the shortfin mako and porbeagle, but is known for a few bony fish as well, such as the halfbeak *Nomorhamphus ebrardtii*. Intrauterine cannibalism is an even more unusual mode of vivipary, in which the largest embryos eat weaker and smaller siblings. This behavior is also most commonly found among sharks, such as the grey nurse shark, but has also been reported for *Nomorhamphus ebrardtii*.

In many species of fish, fins have been modified to allow Internal fertilisation.

Aquarists commonly refer to ovoviviparous and viviparous fish as livebearers.

- Many fish species are hermaphrodites. *Synchronous hermaphrodites* possess both ovaries and testes at the same time. *Sequential hermaphrodites* have both types of tissue in their gonads, with one type being predominant while the fish belongs to the corresponding gender.

Social Behavior

Fish social behavior called 'shoaling' involves a group of fish swimming together. This behavior is a defence mechanism in the sense that there is safety in large numbers, where chances of being eaten by predators are reduced. Shoaling also increases mating and foraging success. Schooling on the other hand, is a behavior within the shoal where fish can be seen performing various manoeuvres in a synchronised manner. The parallel swimming is a form of 'social copying' where fish in the school replicate the direction and velocity of its neighbouring fishes.

Experiments done by D.M. Steven, on the shoaling behavior of fish concluded that during the day, fish had a higher tendency to stay together as a result of a balance between single fish leaving and finding their own direction and the mutual attraction between fishes of the same species. It was found that at night the fish swam noticeably faster however, often singly and in no co-ordination. Groups of two or three could be seen frequently formed although were dispersed after a couple of seconds.

Theoretically, the amount of time that a fish stays together in a shoal should represent their cost of staying instead of leaving. A past laboratory experiment done on cyprinids has established that the time budget for social behavior within a shoal varies proportionally to the quantity of fishes present. This originates from the cost/benefit ratio which changes accordingly with group size, measured by the risk of predation versus food intake. When the cost/benefit ratio is favourable to shoaling behavior then decisions to stay with a group or join one is favourable. Depending on this ratio, fish will correspondingly decide to leave or stay. Thus, shoaling behavior is considered to be driven by an individual fish's constant stream of decisions.

Respiration

Respiration in fish or in that of any organism that lives in the water is very different from that of human beings. Organisms like fish, which live in water, need oxygen to breathe so that their cells can maintain their living state. To perform their respiratory function, fish have specialized organs that help them inhale oxygen dissolved in water.

Respiration in Fish

Respiration in fish takes with the help of gills. Most fish possess gills on either side of their head. Gills are tissues made up of feathery structures called gill filaments that provide a large surface area for gas exchange. A large surface area is crucial for gas exchange in aquatic organisms as water contains very little amount of dissolved oxygen. The filaments in fish gills are arranged in rows in the gill arch. Each filament contains lamellae, which are discs supplied with capillaries. Blood enters and leaves the gills through these small blood vessels. Although gills in fish occupy only a small section of their body, the immense respiratory surface created by the filaments provides the whole organism with an efficient gas exchange.

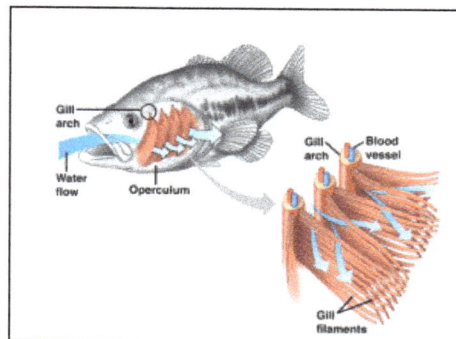

Fish take in oxygen-rich water through their mouths and pump it over their gills. As water passes over the gill filaments, blood inside the capillary network picks up the dissolved oxygen. The circulatory system then transports the oxygen to all body tissues and ultimately to the cells. While picking up carbon dioxide, which is removed from the body through the gills. After the water flows through the gills, it exits the body of the fish through the openings in the sides of the throat or through the operculum, a flap, usually found in bony fish, that covers and protects the fish gills.

Some fish, like sharks and lampreys, possess multiple gill openings. However, bony fish like Rohu, have a single gill opening on each side.

Sensory Systems in Fish

Most fish possess highly developed sense organs. Nearly all daylight fish have color vision that is at least as good as a human's. Many fish also have chemoreceptors that are responsible for extraordinary senses of taste and smell. Although they have ears, many fish may not hear very well. Most fish have sensitive receptors that form the lateral line system, which detects gentle currents and vibrations, and senses the motion of nearby fish and prey. Sharks can sense frequencies in the range of 25 to 50 Hz through their lateral line.

Fish orient themselves using landmarks and may use mental maps based on multiple landmarks or symbols. Fish behavior in mazes reveals that they possess spatial memory and visual discrimination.

Vision

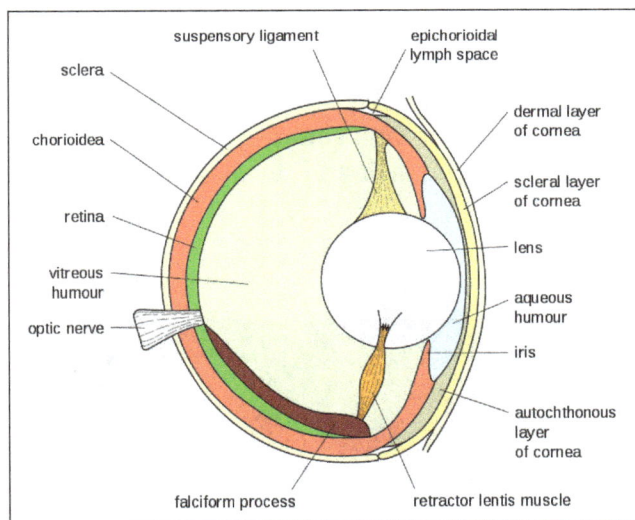

Diagrammatic vertical section through the eye of teleost fish. Fish have a refractive index gradient within the lens which compensates for spherical aberration. Unlike humans, most fish adjust focus by moving the lens closer or further from the retina. Teleosts do so by contracting the retractor lentis muscle.

Vision is an important sensory system for most species of fish. Fish eyes are similar to those of terrestrial vertebrates like birds and mammals, but have a more spherical lens. Their retinas generally have both rod cells and cone cells (for scotopic and photopic vision), and most species have colour vision. Some fish can see ultraviolet and some can see polarized light. Amongst jawless fish, the lamprey has well-developed eyes, while the hagfish has only primitive eyespots. Fish vision shows adaptation to their visual environment, for example deep sea fishes have eyes suited to the dark environment.

Fish and other aquatic animals live in a different light environment than terrestrial species. Water absorbs light so that with increasing depth the amount of light available decreases quickly. The optic properties of water also lead to different wavelengths of light being absorbed to different degrees, for example light of long wavelengths (e.g. red, orange) is absorbed quite quickly compared to light of short wavelengths (blue, violet), though ultraviolet light (even shorter wavelength than blue) is absorbed quite quickly as well. Besides these universal qualities of water, different bodies of water may absorb light of different wavelengths because of salts and other chemicals in the water.

Hearing

Hearing is an important sensory system for most species of fish. Hearing threshold and the ability to localize sound sources are reduced underwater, in which the speed of sound is faster than in air. Underwater hearing is by bone conduction, and localization of sound appears to depend on differences in amplitude detected by bone conduction. As such, aquatic animals such as fish have a more specialized hearing apparatus that is effective underwater.

Fish can sense sound through their lateral lines and their otoliths (ears). Some fishes, such as some species of carp and herring, hear through their swim bladders, which function rather like a hearing aid.

Hearing is well-developed in carp, which have the Weberian organ, three specialized vertebral processes that transfer vibrations in the swim bladder to the inner ear.

Although it is hard to test sharks' hearing, they may have a sharp sense of hearing and can possibly hear prey many miles away. A small opening on each side of their heads (not the spiracle) leads directly into the inner ear through a thin channel. The lateral line shows a similar arrangement, and is open to the environment via a series of openings called lateral line pores. This is a reminder of the common origin of these two vibration- and sound-detecting organs that are grouped together as the acoustico-lateralis system. In bony fish and tetrapods the external opening into the inner ear has been lost.

Current Detection

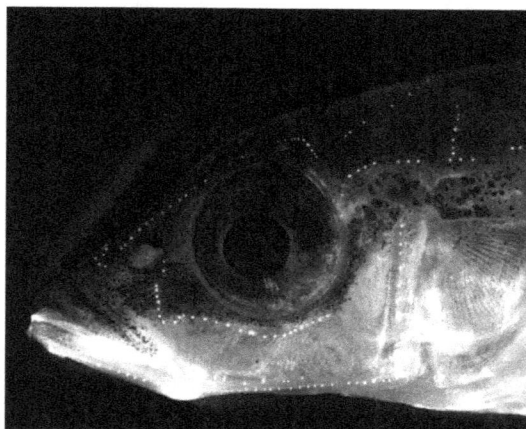

A *three-spined stickleback* with stained neuromasts.

The lateral line in fish and aquatic forms of amphibians is a detection system of water currents, consisting mostly of vortices. The lateral line is also sensitive to low-frequency vibrations. It is used primarily for navigation, hunting, and schooling. The mechanoreceptors are hair cells, the same mechanoreceptors for vestibular sense and hearing. Hair cells in fish are used to detect water movements around their bodies. These hair cells are embedded in a jelly-like protrusion called cupula. The hair cells therefore can not be seen and do not appear on the surface of skin. The receptors of the electrical sense are modified hair cells of the lateral line system.

Fish and some aquatic amphibians detect hydrodynamic stimuli via a lateral line. This system consists of an array of sensors called neuromasts along the length of the fish's body. Neuromasts can be free-standing (superficial neuromasts) or within fluid-filled canals (canal neuromasts). The sensory cells within neuromasts are polarized hair cells contained within a gelatinous cupula. The cupula, and the stereocilia which are the "hairs" of hair cells, are moved by a certain amount depending on the movement of the surrounding water. Afferent nerve fibers are excited or inhibited depending on whether the hair cells they arise from are deflected in the preferred or opposite direction. Lateral line neurons form somatotopic maps within the brain informing the fish of amplitude and direction of flow at different points along the body. These maps are located in the medial octavolateral nucleus (MON) of the medulla and in higher areas such as the torus semicircularis.

Pressure Detection

Pressure detection uses the organ of Weber, a system consisting of three appendages of vertebrae transferring changes in shape of the gas bladder to the middle ear. It can be used to regulate the buoyancy of the fish. Fish like the weather fish and other loaches are also known to respond to low pressure areas but they lack a swim bladder.

Chemoreception

The shape of the hammerhead shark's head may enhance olfaction by spacing the nostrils further apart.

The aquatic equivalent to smelling in air is tasting in water. Many larger catfish have chemoreceptors across their entire bodies, which means they "taste" anything they touch and "smell" any chemicals in the water. "In catfish, gustation plays a primary role in the orientation and location of food".

Salmon have a strong sense of smell. Speculation about whether odours provide homing cues, go back to the 19th century. In 1951, Hasler hypothesised that, once in vicinity of the estuary or entrance to its birth river, salmon may use chemical cues which they can smell, and which are unique to their natal stream, as a mechanism to home onto the entrance of the stream. In 1978, Hasler and his students convincingly showed that the way salmon locate their home rivers with such precision was indeed because they could recognise its characteristic smell. They further demonstrated that the smell of their river becomes imprinted in salmon when they transform into smolts, just before they migrate out to sea. Homecoming salmon can also recognise characteristic smells in tributary streams as they move up the main river. They may also be sensitive to characteristic pheromones given off by juvenile conspecifics. There is evidence that they can "discriminate between two populations of their own species".

Sharks have keen olfactory senses, located in the short duct (which is not fused, unlike bony fish) between the anterior and posterior nasal openings, with some species able to detect as little as one part per million of blood in seawater. Sharks have the ability to determine the direction of a given scent based on the timing of scent detection in each nostril. This is similar to the method mammals use to determine the direction of sound. They are more attracted to the chemicals found in the intestines of many species, and as a result often linger near or in sewage outfalls. Some species, such as nurse sharks, have external barbels that greatly increase their ability to sense prey.

The MHC genes are a group of genes present in many animals and important for the immune system; in general, offspring from parents with differing MHC genes have a stronger immune system. Fish are able to smell some aspect of the MHC genes of potential sex partners and prefer partners with MHC genes different from their own.

Electroreception and Magnetoreception

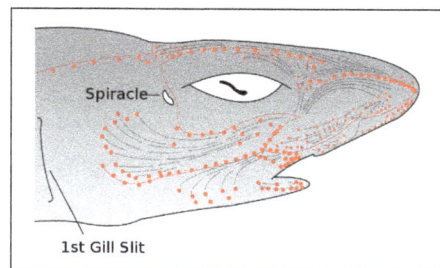

Electromagnetic field receptors (ampullae of Lorenzini) and motion detecting canals in the head of a shark.

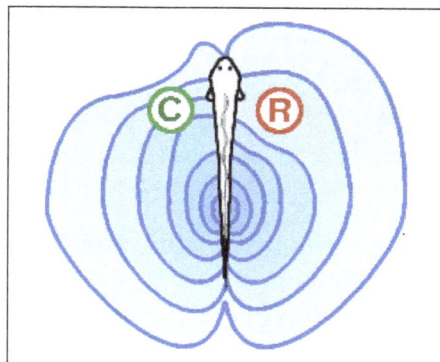

Active electrolocation. Conductive objects concentrate the field and resistive objects spread the field.

Electroreception, or electroception, is the ability to detect electric fields or currents. Some fish, such as catfish and sharks, have organs that detect weak electric potentials on the order of millivolts. Other fish, like the South American electric fishes Gymnotiformes, can produce weak electric currents, which they use in navigation and social communication. In sharks, the ampullae of Lorenzini are electroreceptor organs. They number in the hundreds to thousands. Sharks use the ampullae of Lorenzini to detect the electromagnetic fields that all living things produce. This helps sharks (particularly the hammerhead shark) find prey. The shark has the greatest electrical sensitivity of any animal. Sharks find prey hidden in sand by detecting the electric fields they produce. Ocean currents moving in the magnetic field of the Earth also generate electric fields that sharks can use for orientation and possibly navigation.

Electric field proximity sensing is used by the electric catfish to navigate through muddy waters. These fish make use of spectral changes and amplitude modulation to determine factors such shape, size, distance, velocity, and conductivity. The abilities of the electric fish to communicate and identify sex, age, and hierarchy within the species are also made possible through electric fields. EF gradients as low as 5nV/cm can be found in some saltwater weakly electric fish.

The paddlefish (*Polyodon spathula*) hunts plankton using thousands of tiny passive electroreceptors located on its extended snout, or rostrum. The paddlefish is able to detect electric fields that oscillate at 0.5–20 Hz, and large groups of plankton generate this type of signal.

Electric fishes use an active sensory system to probe the environment and create active electrodynamic imaging.

In 1973, it was shown that Atlantic salmon have conditioned cardiac responses to electric fields with strengths similar to those found in oceans. "This sensitivity might allow a migrating fish to align itself upstream or downstream in an ocean current in the absence of fixed references."

Magnetoception, or magnetoreception, is the ability to detect the direction one is facing based on the Earth's magnetic field. In 1988, researchers found iron, in the form of single domain magnetite, resides in the skulls of sockeye salmon. The quantities present are sufficient for magnetoception.

Fish Navigation

Salmon regularly migrate thousands of miles to and from their breeding grounds.

Salmon spend their early life in rivers, and then swim out to sea where they live their adult lives and gain most of their body mass. After several years wandering huge distances in the ocean where they mature, most surviving salmons return to the same natal rivers to spawn. Usually they return with uncanny precision to the river where they were born: most of them swim up the rivers until they reach the very spawning ground that was their original birthplace.

There are various theories about how this happens. One theory is that there are geomagnetic and chemical cues which the salmon use to guide them back to their birthplace. It is thought that, when they are in the ocean, they use magnetoception related to Earth's magnetic field to orient itself in the ocean and locate the general position of their natal river, and once close to the river, that they use their sense of smell to home in on the river entrance and even their natal spawning ground.

Pain

Experiments done by William Tavolga provide evidence that fish have pain and fear responses. For instance, in Tavolga's experiments, toadfish grunted when electrically shocked and over time they came to grunt at the mere sight of an electrode.

Hooked sailfish.

In 2003, Scottish scientists at the University of Edinburgh and the Roslin Institute concluded that rainbow trout exhibit behaviors often associated with pain in other animals. Bee venom and acetic acid injected into the lips resulted in fish rocking their bodies and rubbing their lips along the sides and floors of their tanks, which the researchers concluded were attempts to relieve pain, similar to what mammals would do. Neurons fired in a pattern resembling human neuronal patterns.

Professor James D. Rose of the University of Wyoming claimed the study was flawed since it did not provide proof that fish possess "conscious awareness, particularly a kind of awareness that is meaningfully like ours". Rose argues that since fish brains are so different from human brains, fish are probably not conscious in the manner humans are, so that reactions similar to human reactions to pain instead have other causes. Rose had published a study a year earlier arguing that fish cannot feel pain because their brains lack a neocortex. However, animal behaviorist Temple Grandin argues that fish could still have consciousness without a neocortex because "different species can use different brain structures and systems to handle the same functions."

Animal welfare advocates raise concerns about the possible suffering of fish caused by angling. Some countries, such as Germany have banned specific types of fishing, and the British RSPCA now prosecutes individuals who are cruel to fish.

Fish Locomotion

Many fishes have a streamlined body and swim freely in open water. Fish locomotion is closely correlated with habitat and ecological niche (the general position of the animal to its environment).

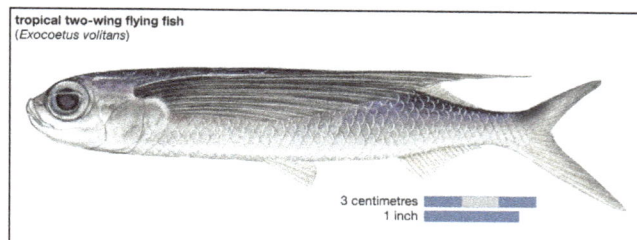

Tropical two-wing flying fish.

Flying fishes, such as the tropical two-wing flying fish (Exocoetus volitans), do not fly in the sense of flapping their wing-sized fins. Their fins do allow them to glide in the air, however, after building up enough speed from swimming to break the surface of the water.

Many fishes in both marine and fresh waters swim at the surface and have mouths adapted to feed best (and sometimes only) at the surface. Often such fishes are long and slender, able to dart at surface insects or at other surface fishes and in turn to dart away from predators; needlefishes, halfbeaks, and topminnows (such as killifish and mosquito fish) are good examples. Oceanic flying fishes escape their predators by gathering speed above the water surface, with the lower lobe of the tail providing thrust in the water. They then glide hundreds of yards on enlarged, winglike pectoral and pelvic fins. South American freshwater flying fishes escape their enemies by jumping and propelling their strongly keeled bodies out of the water.

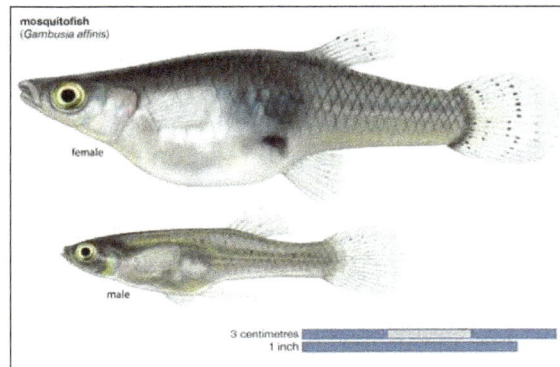

Mosquitofish (Gambusia affinis).

The mosquitofish (Gambusia affinis), a type of topminnow native to the fresh waters of the southeastern United States, bears its young alive.

So-called mid-water swimmers, the most common type of fish, are of many kinds and live in many habitats. The powerful fusiform tunas and the trouts, for example, are adapted for strong, fast swimming, the tunas to capture prey speedily in the open ocean and the trouts to cope with the swift currents of streams and rivers. The trout body form is well adapted to many habitats. Fishes that live in relatively quiet waters such as bays or lake shores or slow rivers usually are not strong, fast swimmers but are capable of short, quick bursts of speed to escape a predator. Many of these fishes have their sides flattened, examples being the sunfish and the freshwater angelfish of aquarists. Fish associated with the bottom or substrate usually are slow swimmers. Open-water plankton-feeding fishes almost always remain fusiform and are capable of rapid, strong movement (for example, sardines and herrings of the open ocean and also many small minnows of streams and lakes).

Bluefin tuna (Thunnus thynnus).

Rainbow trout (Oncorhynchus mykiss).

An aquarium angelfish (Pterophyllum).

Bottom-living fishes are of many kinds and have undergone many types of modification of their body shape and swimming habits. Rays, which evolved from strong-swimming mid-water sharks, usually stay close to the bottom and move by undulating their large pectoral fins. Flounders live in a similar habitat and move over the bottom by undulating the entire body. Many bottom fishes dart from place to place, resting on the bottom between movements, a motion common in gobies. One goby relative, the mudskipper, has taken to living at the edge of pools along the shore of muddy mangrove swamps. It escapes its enemies by flipping rapidly over the mud, out of the water. Some catfishes, synbranchid eels, the so-called climbing perch, and a few other fishes venture out over damp ground to find more promising waters than those that they left. They move by wriggling their bodies, sometimes using strong pectoral fins; most have accessory air-breathing organs. Many bottom-dwelling fishes live in mud holes or rocky crevices. Marine eels and gobies commonly are found in such habitats and for the most part venture far beyond their cavelike homes. Some bottom dwellers, such as the clingfishes (Gobiesocidae), have developed powerful adhesive disks that enable them to remain in place on the substrate in areas such as rocky coasts, where the action of the waves is great.

References

- Vertebrate, entry: newworldencyclopedia.org, Retrieved 21 may 2019

- Storer, Tracy I.; Usinger, R. L.; Stebbins, Robert C.; Nybakken, James W. (1997). General Zoology (sixth ed.). New York: McGraw-Hill. pp. 750–751. ISBN 978-0-07-061780-3

- Sturkie, P. D. (1998). Sturkie's Avian Physiology. 5th Edition. Academic Press, San Diego. ISBN 978-0-12-747605-6. OCLC 162128712

- physiology, birds, animal: basicbiology.net, Retrieved 12 April, 2019

- Nicholls, Henry (10 September 2009). "Mouth to Mouth". Nature. 461(7261): 164–166. doi:10.1038/461164a. PMID 19741680

- Locomotion, fish, animal: britannica.com, Retrieved 19 April, 2019

- Guides. ISBN 978-0-00-713610-0. OCLC 183136093

- Guillaume, Jean; Praxis Publishing; Sadasivam Kaushik; Pierre Bergot; Robert Metailler (2001). Nutrition and Feeding of Fish and Crustaceans. Springer. p. 31. ISBN 978-1-85233-241-9. Retrieved 9 January 2009

- respiration-fish-mechanism, biology: byjus.com, Retrieved 21 July , 2019

3
Invertebrate Physiology

The branch of biology which studies the physiology of animals which neither possess nor develop a vertebral column derived from the notochord is termed as invertebrate physiology. Some of the common invertebrates are insects and worms. This chapter discusses the diverse aspects related to these invertebrates in detail.

Invertebrate is a term used to describe any animal without a backbone or spinal column. The group includes about 97 percent of all animal species; that is, all animals except vertebrates, (subphylum Vertebrata of the phylum Chordata), which have a backbone or spinal column. Invertebrates include simple organisms, such as sponges and flatworms, and more complex animals, such as arthropods and molluscs. Vertebrates include the familiar fish, reptiles, amphibians, birds, and mammals. Since invertebrates include all animals except a certain group, invertebrates form a paraphyletic group.

Common Blue Damselfly (Enallagama cyathigerum), an insect, one of millions of species of invertebrates.

Ubiquitous and filling diverse niches, invertebrates are integral to the ecology, productivity, and harmony of all ecosystems, and central to the extraordinary diversity of life that is so cherished by humans.

Phyla of Invertebrates

The term invertebrate was coined by Jean-Baptiste Lamarck, who divided these animals into two groups, the Insecta and the Vermes. Today, invertebrates are classified into about 30 phyla.

All phyla of animals are invertebrates with the exception that only two of the three subphyla in Phylum Chordata are invertebrates: Urochordata and Cephalochordata. These two, plus all the other known invertebrates, have only one cluster of Hox genes, while the vertebrates have duplicated their original cluster more than once. The largest subphyla in Chordata is Vertebrata.

The exact number of phyla of invertebrates varies according to the taxonomic scheme. For example, some taxonomists recognize a phylum Endoprocta (or Ectoprocta) that exists independently of phylum Bryozoa, but others place both in the single phylum Bryozoa. Some taxonomic schemes recognize Phylum Echiura (spoon worms) and Phylum Pogonophora (beard worms), while other taxonomists assign these the rank of class, with Class Echiura and class Pogoonophora part of the Plylum Annelida. The following is a broad listing of invertebrate phyla:

- Phylum Placozoa (Placozoa),

- Phylum Porifera (sponges),

- Phylum Cnidaria (coral, jellyfish, anemones),

- Phylum Ctenophora (comb jellies),

- Phylum Platyhelminthes (flatworms),

- Phylum Gnathostomulida (jaw worms),

- Phylum Mesozoa (mesozoa),

- Phylum Nemertina (or Phylum Rhynchocoela) (proboscis worms),

- Phylum Gastrotricha (gastrotrichs),

- Phylum Rotifera (rotifers),

- Phylum Nematoda (roundworms),

- Phylum Nematomorpha (horsehair worms),

- Phylum Kinorhyncha (mud dragons, spiny-crown worms),

- Phylum Acanthocephala (acanthocephalans, spiny-headed worms),

- Phylum Loricifera (brush heads),

- Phylum Cycliophora (pandora, cycliophorans),

- Phylum Entoprocta (goblet worms or marine mats),

- Phylum Bryozoa or Phylum Ectoprocta (or Endoprocta) (moss animals or bryozoans),

- Phylum Phoronida (horseshoe worms),

- Phylum Brachiopoda (brachipods, lampshells),

- Phylum Mollusca (molluscs: slugs, snails, squid),

- Phylum Priapulida (priapulid worms),

- Phylum Sipuncula (peanut worms),

- Phylum Annelida (segmented worms: earthworms, ragworms),

- Phylum Echiura (or Class Echiura of Annelida) (spoon worms),

- Phylum Pogonophora (or class Pogonophora of Annelida) (beard worms),

- Phylum Tardigrada (water bears),

- Phylum Onychophora (velvet worms),

- Phylum Arthropoda (insects, spiders, crabs, etc.),

- Phylum Echinodermata (starfish, urchins),

- Phylum Chaetognatha (arrow worms),

- Phylum Hemichordata (acorn worms),

- Phylum Chordata (vertebrates and invertebrates, etc.).
  - Subphylum Urochordata,
  - Subphylum Cephalochordata.

Select Phyla of Invertebrates

The following are descriptions of some well-known invertebrate phyla.

Porifera: Sponges

An elephant ear sponge.

The sponges or poriferans are primitive, sessile, mostly marine, water dwelling filter feeders that pump water through their bodies to filter out particles of food matter. With no true tissues, they lack muscles, nerves, and internal organs. There are over 5,000 modern species of sponges known, and they can be found attached to surfaces anywhere from the intertidal zone to as deep as 8,500 meters (29,000 feet) or further. The fossil record of sponges dates back to the Precambrian era.

Cnidarians: Jellyfish, Corals, Sea Anemones

Cnidaria is a phylum containing some 11,000 species of relatively simple animals found exclusively in aquatic, mostly marine, environments. Cnidarians get their name from cnidocytes, which are specialized cells that carry stinging organelles. The corals, which are important reef-builders, belong here, as do the familiar sea anemones and jellyfish. Cnidarians are highly evident in the fossil records, having first appeared in the Precambrian era.

Platyhelminthes: Flatworms

The flatworms are relatively simple soft-bodied invertebrates. With about 25,000 known species they are the largest phylum of acoelomates. Flatworms are found in marine, freshwater, and even damp terrestrial environments. Most are free-living forms, but many are parasitic on other animals. They include flukes and tapeworms.

Nematoda: Roundworms

The nematodes or roundworms are one of the most common phyla of invertebrates, with over 20,000 different described species, of which over 15,000 are parasitic. They are ubiquitous in freshwater, marine, and terrestrial environments, where they often outnumber other animals in both individual and species counts, and are found in locations as diverse as Antarctica and oceanic trenches. There are a great many parasitic forms, including pathogens in most plants and animals, humans included.

Annelida: Earthworms

The annelids comprise the segmented worms, with about 15,000 modern species, including the well-known earthworms and leeches. They are found in most wet environments, and include many terrestrial, freshwater, and especially marine species (such as the polychaetes), as well as some which are parasitic or mutualistic. They range in length from under a millimeter to over three meters (the seep tube worm Lamellibrachia luymesi).

Echinodermata—Sea Star, Sea Urchins, Sea Cucumbers

Live sand dollar on a beach.

Echinoderms are a phylum of marine invertebrates found at all depths. This phylum appeared in the early Cambrian period and contains about 7,000 living species and 13,000 extinct ones. They include starfish, sea daisies, crinoids, sea urchins, sand dollars, sea cucumbers, and brittle stars. Echinodermata is the largest animal phylum to lack any freshwater or terrestrial representatives.

Mollusca—squid, Snails

The mollusks (American spelling) or molluscs (British spelling) are the large and diverse phylum Mollusca, which includes a variety of familiar animals well-known for their decorative shells or as seafood. These range from tiny snails, clams, and abalone to squid, cuttlefish and the octopus (which is considered the most intelligent invertebrate). There are some 112,000 species within this phylum. The giant squid, which until recently had not been observed alive in its adult form, is the largest invertebrate; although it is possible that the colossal squid is even larger.

Arthropoda—insects, Ticks, Spiders, Grasshoppers, Lobsters, Crabs

Arthropods are the largest phylum of animals and include the insects, arachnids, crustaceans, and others. More than 80 percent of described living animal species are arthropods, with over a million modern species described and a fossil record reaching back to the early Cambrian. Arthropods are common throughout marine, freshwater, terrestrial, and even aerial environments, as well as including various symbiotic and parasitic forms. They range in size from microscopic plankton up to forms several meters long.

Arthropods are characterized by the possession of a segmented body with appendages on each segment. They have a dorsal heart and a ventral nervous system. All arthropods are covered by a hard exoskeleton made of chitin, a polysaccharide, which provides physical protection and resistance to desiccation. Periodically, an arthropod sheds this covering when it molts.

Insect Morphology

Insect morphology is the study and description of the physical form of insects. The terminology used to describe insects is similar to that used for other arthropods due to their shared evolutionary history. Three physical features separate insects from other arthropods: they have a body divided into three regions (head, thorax, and abdomen), have three pairs of legs, and mouthparts located *outside* of the head capsule. It is this position of the mouthparts which divides them from their closest relatives, the non-insect hexapods, which includes Protura, Diplura, and Collembola.

There is enormous variation in body structure amongst insect species. Individuals can range from 0.3 mm (fairyflies) to 30 cm across (great owlet moth);[7] have no eyes or many; well-developed wings or none; and legs modified for running, jumping, swimming, or even digging. These modifications allow insects to occupy almost every ecological niche on the planet, except the deep ocean and the Antarctic.

Anatomy

Insects, like all arthropods, have no interior skeleton; instead, they have an exoskeleton, a hard outer layer made mostly of chitin which protects and supports the body. The insect body is divided into three parts: the head, thorax, and abdomen. The head is specialized for sensory input and food intake; the thorax, which is the anchor point for the legs and wings (if present), is specialized for locomotion; and the abdomen for digestion, respiration, excretion, and reproduction. Although the general function of the three body regions is the same across all insect species, there are major differences in basic structure, with wings, legs, antennae, and mouthparts being highly variable from group to group.

External

Exoskeleton

The insect outer skeleton, the cuticle, is made up of two layers; the epicuticle, which is a thin, waxy, water-resistant outer layer and contains no chitin, and the layer under it called the procuticle. This is chitinous and much thicker than the epicuticle and has two layers, the outer is the exocuticle while the inner is the endocuticle. The tough and flexible endocuticle is built from numerous layers of fibrous chitin and proteins, criss-crossing each other in a sandwich pattern, while the exocuticle is rigid and sclerotized. The exocuticle is greatly reduced in many soft-bodied insects, especially the larval stages (e.g., caterpillars). Chemically, chitin is a long-chain polymer of a N-acetylglucosamine, a derivative of glucose. In its unmodified form, chitin is translucent, pliable, resilient and quite tough. In arthropods, however, it is often modified, becoming embedded in a hardened proteinaceous matrix, which forms much of the exoskeleton. In its pure form, it is leathery, but when encrusted in calcium carbonate, it becomes much harder. The difference between the unmodified and modified forms can be seen by comparing the body wall of a caterpillar (unmodified) to a beetle (modified).

From the embryonic stages itself, a layer of columnar or cuboidal epithelial cells gives rise to the external cuticle and an internal basement membrane. The majority of insect material is held in the endocuticle. The cuticle provides muscular support and acts as a protective shield as the insect develops. However, since it cannot grow, the external sclerotised part of the cuticle is periodically shed in a process called "moulting". As the time for moulting approaches, most of the exocuticle material is reabsorbed. In moulting, first the old cuticle separates from the epidermis (apolysis). Enzymatic moulting fluid is released between the old cuticle and epidermis, which separates the exocuticle by digesting the endocuticle and sequestering its material for the new cuticle. When the new cuticle has formed sufficiently, the epicuticle and reduced exocuticle are shed in ecdysis.

The four principal regions of an insect body segment are: tergum or dorsal, sternum or ventral and the two pleura or laterals. Hardened plates in the exoskeleton are called sclerites, which are subdivisions of the major regions - tergites, sternites and pleurites, for the respective regions tergum, sternum, and pleuron.

Head

The head in most insects is enclosed in a hard, heavily sclerotized, exoskeletal head capsule'. The main exception is in those species whose larvae are not fully sclerotised, mainly some holometabola; but even most unsclerotised or weakly sclerotised larvae tend to have well sclerotised head capsules, for example the larvae of Coleoptera and Hymenoptera. The larvae of Cyclorrhapha however, tend to have hardly any head capsule at all.

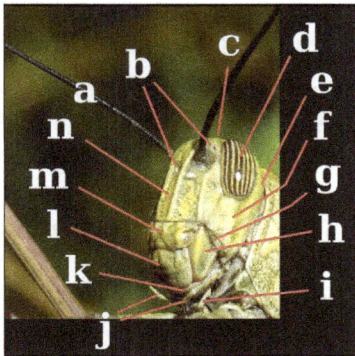

Head of Orthoptera, Acrididae. a:antenna; b:ocelli; c:vertex; d:compound eye; e:occiput; f:gena; g:pleurostoma; h:mandible; i:labial palp; j:maxillary palps; k:maxilla; l:labrum; m:clypeus; n:frons.

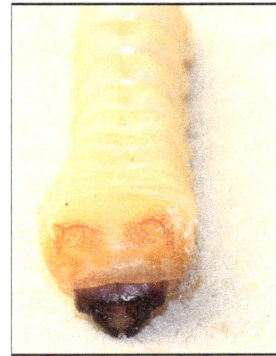

Larva of beetle, family Cerambycidae, showing sclerotised epicranium; rest of body hardly sclerotised.

The head capsule bears most of the main sensory organs, including the antennae, ocelli, and the compound eyes. It also bears the mouthparts. In the adult insect the head capsule is apparently unsegmented, though embryological studies show it to consist of six segments that bear the paired head appendages, including the mouthparts, each pair on a specific segment. Each such pair occupies one segment, though not all segments in modern insects bear any visible appendages.

Larva of Syrphid fly, member of Cyclorrhapha, without epicranium, almost without sclerotisation apart from its jaws.

Of all the insect orders, Orthoptera most conveniently display the greatest variety of features found in the heads of insects, including the sutures and sclerites. Here, the vertex, or the apex (dorsal region), is situated between the compound eyes for insects with hypognathous and opisthognathous heads. In prognathous insects, the vertex is not found between the compound eyes, but rather, where the ocelli are normally found. This is because the primary axis of the head is rotated 90° to become parallel to the primary axis of the body. In some species, this region is modified and assumes a different name.

The ecdysial suture is made of the coronal, frontal, and epicranial sutures plus the ecdysial and cleavage lines, which vary among different species of insects. The ecdysial suture is longitudinally placed on the vertex and separates the epicranial halves of the head to the left and right sides. Depending on the insect, the suture may come in different shapes: like either a Y, U, or V. Those diverging lines that make up the ecdysial suture are called the frontal or frontogenal sutures. Not all species of insects have frontal sutures, but in those that do, the sutures split open during ecdysis, which helps provide an opening for the new instar to emerge from the integument.

The frons is that part of the head capsule that lies ventrad or anteriad of the vertex. The frons varies in size relative to the insect, and in many species the definition of its borders is arbitrary, even in some insect taxa that have well-defined head capsules. In most species, though, the frons is bordered at its anterior by the frontoclypeal or epistomal sulcus above the clypeus. Laterally it is limited by the fronto-genal sulcus, if present, and the boundary with the vertex, by the ecdysial cleavage line, if it is visible. If there is a median ocellus, it generally is on the frons, though in some insects such as many Hymenoptera, all three ocelli appear on the vertex. A more formal definition is that it is the sclerite from which the pharyngeal dilator muscles arise, but in many contexts that too, is not helpful. In the anatomy of some taxa, such as many Cicadomorpha, the front of the head is fairly clearly distinguished and tends to be broad and sub-vertical; that median area commonly is taken to be the frons.

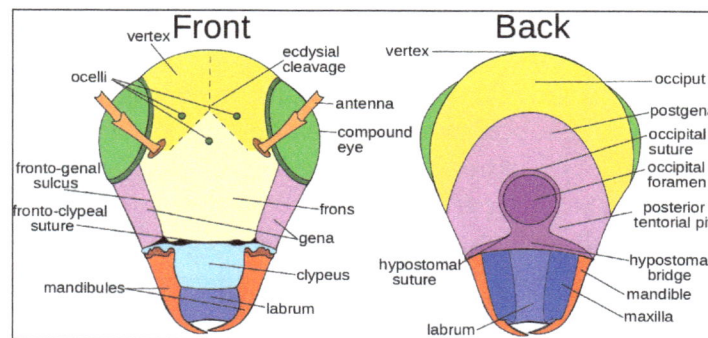

The clypeus is a sclerite between the face and labrum, which is dorsally separated from the frons by the frontoclypeal suture in primitive insects. The clypeogenal suture laterally demarcates the clypeus, with the clypeus ventrally separated from the labrum by the clypeolabral suture. The clypeus differs in shape and size, such as species of Lepidoptera with a large clypeus with elongated mouthparts. The cheek or gena forms the sclerotized area on each side of the head below the compound eyes extending to the gular suture. Like many of the other parts making up the insect's head, the gena varies among species, with its boundaries difficult to establish. For example, in dragonflies and damselflies, it is between the compound eyes, clypeus, and mouthparts. The postgena is the area immediately posteriad, or posterior or lower on the gena of pterygote insects, and forms the lateral and ventral parts of the occipital arch. The occipital arch is a narrow band forming the posterior edge of the head capsule arching dorsally over the foramen. The subgenal area is usually narrow, located above the mouthparts; this area also includes the hypostoma and pleurostoma. The vertex extends anteriorly above the bases of the antennae as a prominent, pointed, concave rostrum. The posterior wall of the head capsule is penetrated by a large aperture, the foramen. Through it pass the organ systems, such as nerve cord, esophagus, salivary ducts, and musculature, connecting the head with the thorax.

On the posterior aspect of the head are the occiput, postgena, occipital foramen, posterior tentorial pit, gula, postgenal bridge, hypostomal suture and bridge, and the mandibles, labium, and maxilla. The occipital suture is well founded in species of Orthoptera, but not so much in other orders. Where found, the occipital suture is the arched, horseshoe-shaped groove on the back of the head, ending at the posterior of each mandible. The postoccipital suture is a landmark on the posterior surface of the head, and is typically near the occipital foremen. In pterygotes, the postocciput forms the extreme posterior, often U-shaped, which forms the rim of the head extending to the postoccipital suture. In pterygotes, such as those of Orthoptera, the occipital foramen and the mouth are not separated. The three types of occipital closures, or points under the occipital foramen that separate the two lower halves of the postgena, are: the hypostomal bridge, the postgenal bridge, and the gula. The hypostomal bridge is usually found in insects with hypognathous orientation. The postgenal bridge is found in the adults of species of higher Diptera and aculeate Hymenoptera, while the gula is found on some Coleoptera, Neuroptera, and Isoptera, which typically display prognathous-oriented mouthparts.

Compound Eyes and Ocelli

Most insects have one pair of large, prominent compound eyes composed of units called ommatidia (ommatidium, singular), possibly up to 30,000 in a single compound eye of, for example, large dragonflies. This type of eye gives less resolution than eyes found in vertebrates, but it gives acute perception of movement and usually possesses UV- and green sensitiity and may have additional sensitivity peaks in other regions of the visual spectrum. Often an ability to detect the E-vector of polarized light exists polarization of light. There can also be an additional two or three ocelli, which help detect low light or small changes in light intensity. The image perceived is a combination of inputs from the numerous ommatidia, located on a convex surface, thus pointing in slightly different directions. Compared with simple eyes, compound eyes possess very large view angles and better acuity than the insect's dorsal ocelli, but some stemmatal (= larval eyes), for example those of sawfly larvae (Tenthredinidae) with an acuity 4 degrees and very high polarization sensitivity, match the performance of compound eyes.

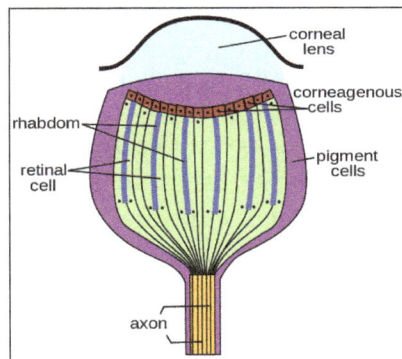

Ocellus cross-section.

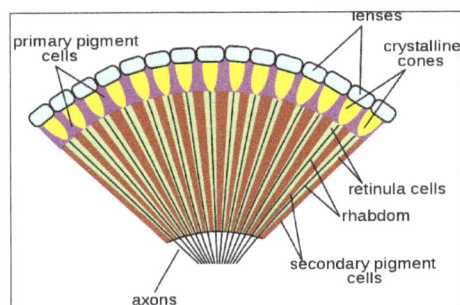

Compound eye cross-section.

Because the individual lenses are so small, the effects of diffraction impose a limit on the possible resolution that can be obtained (assuming they do not function as phased arrays). This can only be countered by increasing lens size and number. To see with a resolution comparable to our simple eyes, humans would require compound eyes that would each reach the size of their heads. Compound eyes fall into two groups: apposition eyes, which form multiple inverted images, and superposition eyes, which form a single erect image. Compound eyes grow at their margins by the addition of new ommatidia.

Antennae

Closeup of a fire ant, showing fine sensory hairs on antennae.

Antennae, sometimes called "feelers", are flexible appendages located on the insect's head which are used for sensing the environment. Insects *are* able to feel with their antennae because of the fine hairs (setae) that cover them. However, touch is not the only thing that antennae can detect; numerous tiny sensory structures on the antennae allow insects to sense smells, temperature, humidity, pressure, and even potentially sense themselves in space. Some insects, including bees and some groups of flies can also detect sound with their antennae.

The number of segments in an antenna varies considerably amongst insects, with higher flies having only 3-6 segments, while adult cockroaches can have over 140. The general shape of the antennae is also quite variable, but the first segment (the one attached to the head) is always called the scape, and the second segment is called the pedicel. The remaining antennal segments or flagellomeres are called the flagellum.

General insect antenna types are shown below:

Types of insect antennae						
Aristate	Capitate	Clavate	Filiform	Flabellate	Geniculate	Setaceous
Lamellate	Moniliform	Pectinate	Plumose	Serrate	Stylate	

Mouthparts

The insect mouthparts consist of the maxilla, labium, and in some species, the mandibles. The labrum is a simple, fused sclerite, often called the upper lip, and moves longitudinally, which is hinged to the clypeus. The mandibles (jaws) are a highly sclerotized pair of structures that move at right angles to the body, used for biting, chewing, and severing food. The maxillae are paired structures that can also move at right angles to the body and possess segmented palps. The labium (lower lip) is the fused structure that moves longitudinally and possesses a pair of segmented palps.

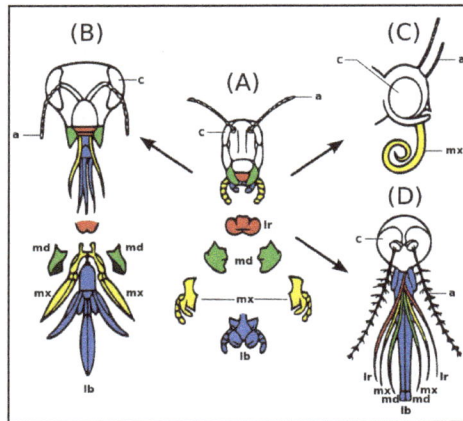

The development of insect mouthparts from the primitive chewing mouthparts of a grasshopper in the centre (A), to the lapping type (B) of a bee, the siphoning type (C) of a butterfly and the sucking type (D) of a female mosquito.
Legend: a - antennae c - compound eye lb - labium lr - labrum md - mandibles mx - maxillae

The mouthparts, along with the rest of the head, can be articulated in at least three different positions: prognathous, opisthognathous, and hypognathous. In species with prognathous articulation, the head is positioned vertically aligned with the body, such as species of Formicidae; while in a hypognathous type, the head is aligned horizontally adjacent to the body. An opisthognathous head is positioned diagonally, such as species of Blattodea and some Coleoptera. The mouthparts vary greatly between insects of different orders, but the two main functional groups are mandibulate and haustellate. Haustellate mouthparts are used for sucking liquids and can be further classified by the presence of stylets, which include piercing-sucking, sponging, and siphoning. The stylets are needle-like projections used to penetrate plant and animal tissues. The stylets and the feeding tube form the modified mandibles, maxilla, and hypopharynx.

- Mandibulate mouthparts, among the most common in insects, are used for biting and grinding solid foods.

- Piercing-sucking mouthparts have stylets, and are used to penetrate solid tissue and then suck up liquid food.

- Sponging mouthparts are used to sponge and suck liquids, and lack stylets (e.g. most Diptera).

- Siphoning mouthparts lack stylets and are used to suck liquids, and are commonly found among species of Lepidoptera.

Mandibular mouthparts are found in species of Odonata, adult Neuroptera, Coleoptera, Hymenoptera, Blattodea, Orthoptera, and Lepidoptera. However, most adult Lepidoptera have siphoning mouthparts, while their larvae (commonly called caterpillars) have mandibles.

Mandibulate

The labrum is a broad lobe forming the roof of the preoral cavity, suspended from the clypeus in front of the mouth and forming the upper lip. On its inner side, it is membranous and may be produced into a median lobe, the epipharynx, bearing some sensilla. The labrum is raised away from the mandibles by two muscles arising in the head and inserted medially into the anterior margin of the labrum. It is closed against the mandibles in part by two muscles arising in the head and inserted on the posterior lateral margins on two small sclerites, the tormae, and, at least in some insects, by a resilin spring in the cuticle at the junction of the labrum with the clypeus. Until recently, the labrum generally was considered to be associated with first head segment. However, recent studies of the embryology, gene expression, and nerve supply to the labrum show it is innervated by the tritocerebrum of the brain, which is the fused ganglia of the third head segment. This is formed from fusion of parts of a pair of ancestral appendages found on the third head segment, showing their relationship. Its ventral, or inner, surface is usually membranous and forms the lobe-like epipharynx, which bears mechanosensilla and chemosensilla.

Chewing insects have two mandibles, one on each side of the head. The mandibles are positioned between the labrum and maxillae. The mandibles cut and crush food, and may be used for defense; generally, they have an apical cutting edge, and the more basal molar area grinds the food. They can be extremely hard (around 3 on Mohs, or an indentation hardness of about $30 \, kg/mm^2$); thus, many termites and beetles have no physical difficulty in boring through foils made from such common metals as copper, lead, tin, and zinc. The cutting edges are typically strengthened by the addition of zinc, manganese, or rarely, iron, in amounts up to about 4% of the dry weight. They are typically the largest mouthparts of chewing insects, being used to masticate (cut, tear, crush, chew) food items. They open outwards (to the sides of the head) and come together medially. In carnivorous, chewing insects, the mandibles can be modified to be more knife-like, whereas in herbivorous chewing insects, they are more typically broad and flat on their opposing faces (e.g., caterpillars). In male stag beetles, the mandibles are modified to such an extent as to not serve any feeding function, but are instead are used to defend mating sites from other males. In ants, the mandibles also serve a defensive function (particularly in soldier castes). In bull ants, the mandibles are elongated and toothed, used as hunting (and defensive) appendages.

Situated beneath the mandibles, paired maxillae manipulate food during mastication. Maxillae can have hairs and "teeth" along their inner margins. At the outer margin, the galea is a cupped or scoop-like structure, which sits over the outer edge of the labium. They also have palps, which are used to sense the characteristics of potential foods. The maxillae occupy a lateral position, one on each side of the head behind the mandibles. The proximal part of the maxilla consists of a basal cardo, which has a single articulation with the head, and a flat plate, the stipes, hinged to the cardo. Both cardo and stipes are loosely joined to the head by membrane so they are capable of movement. Distally on the stipes are two lobes, an inner lacinea and an outer galea, one or both of which may be absent. More laterally on the stipes is a jointed, leglike palp made up of a number of segments; in Orthoptera there are five. Anterior and posterior rotator muscles are inserted on the cardo, and ventral adductor muscles arising on the tentorium are inserted on both cardo and stipes. Arising in the stipes are flexor muscles of lacinea and galea and another lacineal flexor arises in the cranium, but neither the lacinea nor the galea has an extensor muscle. The palp has levator and depressor muscles arising in the stipes, and each segment of the palp has a single muscle causing flexion of the next segment.

In mandibulate mouthparts, the labium is a quadrupedal structure, although it is formed from two fused secondary maxillae. It can be described as the floor of the mouth. With the maxillae, it assists with manipulation of food during mastication or chewing or, in the unusual case of the dragonfly nymph, extends out to snatch prey back to the head, where the mandibles can eat it. The labium is similar in structure to the maxilla, but with the appendages of the two sides fused by the midline, so they come to form a median plate. The basal part of the labium, equivalent to the maxillary cardines and possibly including a part of the sternum of the labial segment, is called the postmentum. This may be subdivided into a proximal submentum and a distal mentum. Distal to the postmentum, and equivalent to the fused maxillary stipites, is the prementum. The prementum closes the preoral cavity from behind. Terminally, it bears four lobes, two inner glossae, and two outer paraglossae, which are collectively known as the ligula. One or both pairs of lobes may be absent or they may be fused to form a single median process. A palp arises from each side of the prementum, often being three-segmented.

The hypopharynx is a median lobe immediately behind the mouth, projecting forwards from the back of the preoral cavity; it is a lobe of uncertain origin, but perhaps associated with the mandibular segment; in apterygotes, earwigs, and nymphal mayflies, the hypopharynx bears a pair of lateral lobes, the superlinguae (singular: superlingua). It divides the cavity into a dorsal food pouch, or cibarium, and a ventral salivarium into which the salivary duct opens. It is commonly found fused to the libium. Most of the hypopharynx is membranous, but the adoral face is sclerotized distally, and proximally contains a pair of suspensory sclerites extending upwards to end in the lateral wall of the stomodeum. Muscles arising on the frons are inserted into these sclerites, which distally are hinged to a pair of lingual sclerites. These, in turn, have inserted into them antagonistic pairs of muscles arising on the tentorium and labium. The various muscles serve to swing the hypopharynx forwards and back, and in the cockroach, two more muscles run across the hypopharynx and dilate the salivary orifice and expand the salivarium.

Examples of Mandibles:

Piercing-sucking

Mouthparts can have multiple functions. Some insects combine piercing parts along with sponging ones which are then used to pierce through tissues of plants and animals. Female mosquitoes feed on blood (hemophagous) making them disease vectors. The mosquito mouthparts consist of the proboscis, paired mandibles and maxillae. The maxillae form needle-like structures, called stylets, which are enclosed by the labium. When mosquito bites, maxillae penetrate the skin and anchor the mouthparts, thus allowing other parts to be inserted. The sheath-like labium slides back, and the remaining mouthparts pass through its tip and into the tissue. Then, through the hypopharynx, the mosquito injects saliva, which contains anticoagulants to stop the blood from clotting. And finally, the labrum (upper lip) is used to suck up the blood. Species of the genus *Anopheles* are characterized by their long palpi (two parts with widening end), almost reaching the end of labrum.

Examples of Piercing Mouthparts:

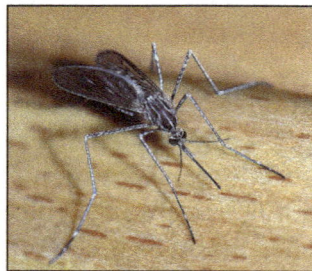

Mosquito

Horsefly (female)

Aedes aegypti

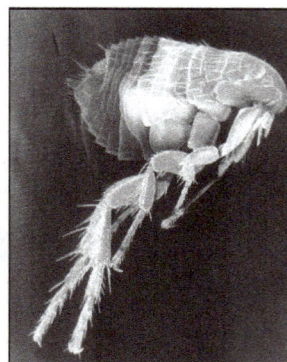

Flea

Siphoning

The proboscis is formed from maxillary galeae and is adaption found in some insects for sucking. The muscles of the cibarium or pharynx are strongly developed and form the pump. In Hemiptera and many Diptera, which feed on fluids within plants or animals, some components of the mouthparts are modified for piercing, and the elongated structures are called stylets. The combined tubular structures are referred to as the proboscis, although specialized terminology is used in some groups.

In species of Lepidoptera, it consists of two tubes held together by hooks and separable for cleaning. Each tube is inwardly concave, thus forming a central tube through which moisture is sucked. Suction is effected through the contraction and expansion of a sac in the head. The proboscis is coiled under the head when the insect is at rest, and is extended only when feeding. The maxillary palpi are reduced or even vestigial. They are conspicuous and five-segmented in some of the more basal families, and are often folded. The shape and dimensions of the proboscis have evolved to give different species wider and therefore more advantageous diets. There is an allometric scaling relationship between body mass of Lepidoptera and length of proboscis from which an interesting adaptive departure is the unusually long-tongued hawk moth *Xanthopan morganii praedicta*. Charles Darwin predicted the existence and proboscis length of this moth before its discovery based on his knowledge of the long-spurred Madagascan star orchid *Angraecum sesquipedale*.

Examples of Siphoning Mouthparts:

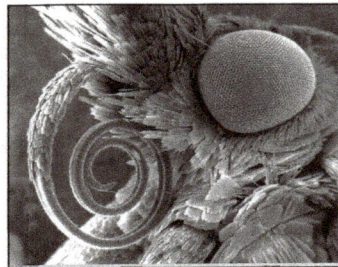

Sponging

The mouthparts of insects that feed on fluids are modified in various ways to form a tube through which liquid can be drawn into the mouth and usually another through which saliva passes.

The muscles of the cibarium or pharynx are strongly developed to form a pump. In nonbiting flies, the mandibles are absent and other structures are reduced; the labial palps have become modified to form the labellum, and the maxillary palps are present, although sometimes short. In Brachycera, the labellum is especially prominent and used for sponging liquid or semiliquid food. The labella are a complex structure consisting of many grooves, called pseudotracheae, which sop up liquids. Salivary secretions from the labella assist in dissolving and collecting food particles so they can be more easily taken up by the pseudotracheae; this is thought to occur by capillary action. The liquid food is then drawn up from the pseudotracheae through the food channel into the esophagus.

The mouthparts of bees are of a chewing and lapping-sucking type. Lapping is a mode of feeding in which liquid or semiliquid food adhering to a protrusible organ, or "tongue", is transferred from substrate to mouth. In the honey bee (Hymenoptera: Apidae: *Apis mellifera*), the elongated and fused labial glossae form a hairy tongue, which is surrounded by the maxillary galeae and the labial palps to form a tubular proboscis containing a food canal. In feeding, the tongue is dipped into the nectar or honey, which adheres to the hairs, and then is retracted so the adhering liquid is carried into the space between the galeae and labial palps. This back-and-forth glossal movement occurs repeatedly. Movement of liquid to the mouth apparently results from the action of the cibarial pump, facilitated by each retraction of the tongue pushing liquid up the food canal.

Examples of Sponging Mouthparts:

Thorax

The insect thorax has three segments: the prothorax, mesothorax, and metathorax. The anterior segment, closest to the head, is the prothorax; its major features are the first pair of legs and the pronotum. The middle segment is the mesothorax; its major features are the second pair of legs and the anterior wings, if any. The third, the posterior, thoracic segment, abutting the abdomen, is the metathorax, which bears the third pair of legs and the posterior wings. Each segment is dilineated by an intersegmental suture. Each segment has four basic regions. The dorsal surface is called the tergum (or notum, to distinguish it from the abdominal terga). The two lateral regions are called the pleura (singular: pleuron), and the ventral aspect is called the sternum. In turn, the notum of the prothorax is called the pronotum, the notum for the mesothorax is called the mesonotum and the notum for the metathorax is called the metanotum. Continuing with this logic, there is also the mesopleura and metapleura, as well as the mesosternum and metasternum.

The tergal plates of the thorax are simple structures in apterygotes and in many immature insects, but are variously modified in winged adults. The pterothoracic nota each have two main divisions: the anterior, wing-bearing alinotum and the posterior, phragma-bearing postnotum. Phragmata (singular: phragma) are plate-like apodemes that extend inwards below the antecostal sutures, marking the primary intersegmental folds between segments; phragmata provide attachment for the longitudinal flight muscles. Each alinotum (sometimes confusingly referred to as a "notum") may be traversed by sutures that mark the position of internal strengthening ridges, and commonly divides the plate into three areas: the anterior prescutum, the scutum, and the smaller posterior scutellum. The lateral pleural sclerites are believed to be derived from the subcoxal segment of the ancestral insect leg. These sclerites may be separate, as in silverfish, or fused into an almost continuous sclerotic area, as in most winged insects.

Prothorax

The pronotum of the prothorax may be simple in structure and small in comparison with the other nota, but in beetles, mantids, many bugs, and some Orthoptera, the pronotum is expanded, and in cockroaches, it forms a shield that covers part of the head and mesothorax.

Pterothorax

Because the mesothorax and metathorax hold the wings, they have a combined name called the pterothorax (pteron = wing). The forewing, which goes by different names in different orders (e.g., the tegmina in Orthoptera and elytra in Coleoptera), arises between the mesonotum and the mesopleuron, and the hindwing articulates between the metanotum and metapleuron. The legs arise from the mesopleuron and metapleura. The mesothorax and metathorax each have a pleural suture (mesopleural and metapleural sutures) that runs from the wing base to the coxa of the leg. The sclerite anterior to the pleural suture is called the episternum (serially, the mesepisternum and metepisternum). The sclerite posterior to the suture is called the epimiron (serially, the mesepimiron and metepimiron). Spiracles, the external organs of the respiratory system, are found on the pterothorax, usually one between the pro- and mesopleoron, as well as one between the meso- and metapleuron.

The ventral view or sternum follows the same convention, with the prosternum under the prothorax, the mesosternum under the mesothorax and the metasternum under the metathorax. The notum, pleura, and sternum of each segment have a variety of different sclerites and sutures, varying greatly from order to order.

Wings

Most phylogenetically advanced insects have two pairs of wings located on the second and third thoracic segments. Insects are the only invertebrates to have developed flight capability, and this has played an important part in their success. Insect flight is not very well understood, relying heavily on turbulent aerodynamic effects. The primitive insect groups use muscles that act directly on the wing structure. The more advanced groups making up the Neoptera have foldable wings, and their muscles act on the thorax wall and power the wings indirectly. These muscles are able to contract multiple times for each single nerve impulse, allowing the wings to beat faster than would ordinarily be possible.

Insect flight can be extremely fast, maneuverable, and versatile, possibly due to the changing shape, extraordinary control, and variable motion of the insect wing. Insect orders use different flight mechanisms; for example, the flight of a butterfly can be explained using steady-state, non-transitory aerodynamics, and thin airfoil theory.

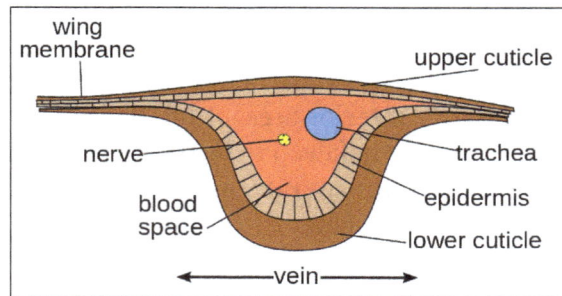

Internal

Each of the wings consists of a thin membrane supported by a system of veins. The membrane is formed by two layers of integument closely apposed, while the veins are formed where the two layers remain separate and the cuticle may be thicker and more heavily sclerotized. Within each of the major veins is a nerve and a trachea, and, since the cavities of the veins are connected with the hemocoel, hemolymph can flow into the wings. Also, the wing lumen, being an extension of the hemocoel, contains the tracheae, nerves, and hemolymph. As the wing develops, the dorsal and ventral integumental layers become closely apposed over most of their area, forming the wing membrane. The remaining areas form channels, the future veins, in which the nerves and tracheae may occur. The cuticle surrounding the veins becomes thickened and more heavily sclerotized to provide strength and rigidity to the wing. Hairs of two types may occur on the wings: microtrichia, which are small and irregularly scattered, and macrotrichia, which are larger, socketed, and may be restricted to veins. The scales of Lepidoptera and Trichoptera are highly modified macrotrichia.

Veins

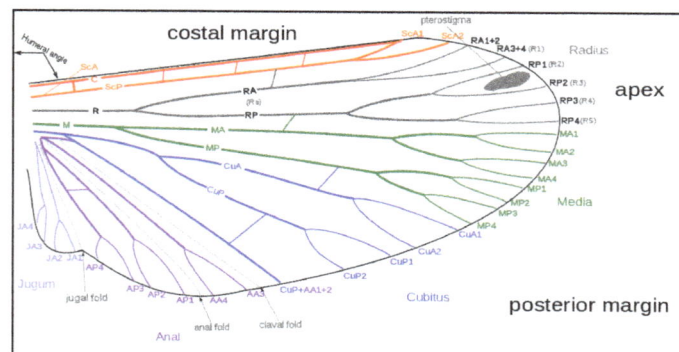

Venation of insect wings, based on the Comstock-Needham system.

In some very small insects, the venation may be greatly reduced. In chalcid wasps, for instance, only the subcosta and part of the radius are present. Conversely, an increase in venation may occur by the branching of existing veins to produce accessory veins or by the development of additional, intercalary veins between the original ones, as in the wings of Orthoptera (grasshoppers and crickets). Large numbers of cross-veins are present in some insects, and they may form a reticulum as in the wings of Odonata (dragonflies and damselflies) and at the base of the forewings of Tettigonioidea and Acridoidea (katydids and grasshoppers, respectively).

The archedictyon is the name given to a hypothetical scheme of wing venation proposed for the very first winged insect. It is based on a combination of speculation and fossil data. Since all winged insects are believed to have evolved from a common ancestor, the archediction represents the "template" that has been modified (and streamlined) by natural selection for 200 million years. According to current dogma, the archedictyon contained six to eight longitudinal veins. These veins (and their branches) are named according to a system devised by John Comstock and George Needham—the Comstock-Needham system:

- Costa (C) - The leading edge of the wing.

- Subcosta (Sc) - Second longitudinal vein (behind the costa), typically unbranched.

- Radius (R) - Third longitudinal vein, one to five branches reach the wing margin.

- Media (M) - Fourth longitudinal vein, one to four branches reach the wing margin.

- Cubitus (Cu) - Fifth longitudinal vein, one to three branches reach the wing margin.

- Anal veins (A1, A2, A3) - Unbranched veins behind the cubitus.

The costa (C) is the leading marginal vein on most insects, although a small vein, the precosta, is sometimes found above the costa. In almost all extant insects, the precosta is fused with the costa; the costa rarely ever branches because it is at the leading edge, which is associated at its base with the humeral plate. The trachea of the costal vein is perhaps a branch of the subcostal trachea. Located after the costa is the third vein, the subcosta, which branches into two separate veins: the anterior and posterior. The base of the subcosta is associated with the distal end of the neck of the first axillary. The fourth vein is the radius, which is branched into five separate veins. The radius is generally the strongest vein of the wing. Toward the middle of the wing, it forks into a first undivided branch (R1) and a second branch, called the radial sector (Ra), which subdivides dichotomously into four distal branches (R2, R3, R4, R5). Basally, the radius is flexibly united with the anterior end of the second axillary (2Ax).

The fifth vein of the wing is the media. In the archetype pattern (A), the media forks into two main branches, a media anterior (MA), which divides into two distal branches (MA1, MA2), and a median sector, or media posterior (MP), which has four terminal branches (M1, M2, M3, M4). In most modern insects, the media anterior has been lost, and the usual "media" is the four-branched media posterior with the common basal stem. In the Ephemerida, according to present interpretations of the wing venation, both branches of the media are retained, while in Odonata, the persisting media is the primitive anterior branch. The stem of the media is often united with the radius, but when it occurs as a distinct vein, its base is associated with the distal median plate (m') or is continuously sclerotized with the latter. The cubitus, the sixth vein of the wing, is primarily two-branched. The primary forking takes place near the base of the wing, forming the two principal branches (Cu1, Cu2). The anterior branch may break up into a number of secondary branches, but commonly it forks into two distal branches. The second branch of the cubitus (Cu2) in Hymenoptera, Trichoptera, and Lepidoptera, was mistaken by Comstock and Needham for the first anal. Proximally, the main stem of the cubitus is associated with the distal median plate (m') of the wing base.

The postcubitus (Pcu) is the first anal of the Comstock and Needham system. The postcubitus, however, has the status of an independent wing vein and should be recognized as such. In nymphal wings, its trachea arises between the cubital trachea and the group of vannal tracheae. In the mature wings of more generalized insects, the postcubitus is always associated proximally with the cubitus, and is never intimately connected with the flexor sclerite (3Ax) of the wing base. In Neuroptera, Mecoptera, and Trichoptera, the postcubitus may be more closely associated with the vannal veins, but its base is always free from the latter. The postcubitus is usually unbranched; primitively, it is two-branched. The vannal veins (lV to nV) are the anal veins immediately associated with the third axillary, and which are directly affected by the movement of this sclerite that brings about the flexion of the wings. In number, the vannal veins vary from one to 12, according to the expansion of the vannal area of the wing. The vannal tracheae usually arise from a common tracheal stem in nymphal insects, and the veins are regarded as branches of a single anal vein. Distally, the vannal veins are either simple or branched. The jugal vein (J) of the jugal lobe of the wing is often occupied by a network of irregular veins, or it may be entirely membranous; sometimes it contains one or two distinct, small veins, the first jugal vein, or vena arcuata, and the second jugal vein, or vena cardinalis (2J).

- C-Sc cross-veins - run between the costa and subcosta.

- R cross-veins - run between adjacent branches of the radius.

- R-M cross-veins - run between the radius and media.

- M-Cu cross-veins - run between the media and cubitus.

All the veins of the wing are subject to secondary forking and to union by cross-veins. In some orders of insects, the cross-veins are so numerous, the whole venational pattern becomes a close network of branching veins and cross-veins. Ordinarily, however, a definite number of cross-veins having specific locations occurs. The more constant cross-veins are the humeral cross-vein (h) between the costa and subcosta, the radial cross-vein (r) between R and the first fork of Rs, the sectorial cross-vein (s) between the two forks of R8, the median cross-vein (m-m) between M2 and M3, and the mediocubital cross-vein (m-cu) between the media and the cubitus.

The veins of insect wings are characterized by a convex-concave placement, such as those seen in mayflies (i.e., concave is "down" and convex is "up"), which alternate regularly and by their branching; whenever a vein forks there is always an interpolated vein of the opposite position between the two branches. The concave vein will fork into two concave veins (with the interpolated vein being convex) and the regular alteration of the veins is preserved. The veins of the wing appear to fall into an undulating pattern according to whether they have a tendency to fold up or down when the wing is relaxed. The basal shafts of the veins are convex, but each vein forks distally into an anterior convex branch and a posterior concave branch. Thus, the costa and subcosta are regarded as convex and concave branches of a primary first vein, Rs is the concave branch of the radius, posterior media the concave branch of the media, Cu1 and Cu2 are respectively convex and concave, while the primitive postcubitus and the first vannal have each an anterior convex branch and a posterior concave branch. The convex or concave nature of the veins has been used as evidence in determining the identities of the persisting distal branches of the veins of modern insects, but it has not been demonstrated to be consistent for all wings.

Fields

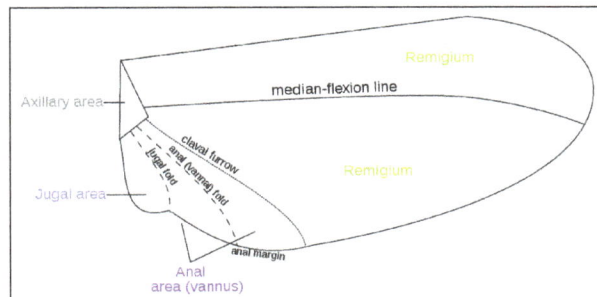

Wing areas are delimited and subdivided by fold lines, along which the wings can fold, and flexion lines, which flex during flight. Between the flexion and the fold lines, the fundamental distinction is often blurred, as fold lines may permit some flexibility or vice versa. Two constants, found in nearly all insect wings, are the claval (a flexion line) and jugal folds (or fold line), forming variable and unsatisfactory boundaries. Wing foldings can very complicated, with transverse folding occurring in the hindwings of Dermaptera and Coleoptera, and in some insects, the anal area can be folded like a fan. The four different fields found on insect wings are:

- Remigium,
- Anal area (vannus),
- Jugal area,
- Axillary area,
- Alula.

Most veins and cross-veins occur in the anterior area of the remigium, which is responsible for most of the flight, powered by the thoracic muscles. The posterior portion of the remigium is sometimes called the clavus the two other posterior fields are the anal and jugal areas. When the vannal fold has the usual position anterior to the group of anal veins, the remigium contains the costal, subcostal, radial, medial, cubital, and postcubital veins. In the flexed wing, the remigium turns posteriorly on the flexible basal connection of the radius with the second axillary, and the base of the mediocubital field is folded medially on the axillary region along the plica basalis (bf) between the median plates (m, m') of the wing base.

The vannus is bordered by the vannal fold, which typically occurs between the postcubitus and the first vannal vein. In Orthoptera, it usually has this position. In the forewing of Blattidae, however, the only fold in this part of the wing lies immediately before the postcubitus. In Plecoptera, the vannal fold is posterior to the postcubitus, but proximally it crosses the base of the first vannal vein. In the cicada, the vannal fold lies immediately behind the first vannal vein (lV). These small variations in the actual position of the vannal fold, however, do not affect the unity of action of the vannal veins, controlled by the flexor sclerite (3Ax), in the flexion of the wing. In the hindwings of most Orthoptera, a secondary vena dividens forms a rib in the vannal fold. The vannus is usually triangular in shape, and its veins typically spread out from the third axillary like the ribs of a fan. Some of the vannal veins may be branched, and secondary veins may alternate with the primary veins. The vannal region is usually best developed in the hindwing, in which it may be enlarged to form a sustaining surface, as in Plecoptera and Orthoptera. The great fan-like expansions of the hindwings of Acrididae are clearly the vannal regions, since their veins are all supported on the third axillary sclerites on the wing bases, though Martynov (1925) ascribes most of the fan areas in Acrididae to the jugal regions of the wings. The true jugum of the acridid wing is represented only by the small membrane (Ju) mesad of the last vannal vein. The jugum is more highly developed in some other Orthoptera, as in the Mantidae. In most of the higher insects with narrow wings, the vannus becomes reduced, and the vannal fold is lost, but even in such cases, the flexed wing may bend along a line between the postcubitus and the first vannal vein.

The jugal region, or neala, is a region of the wing that is usually a small membranous area proximal to the base of the vannus strengthened by a few small, irregular vein-like thickenings; but when well developed, it is a distinct section of the wing and may contain one or two jugal veins. When the jugal area of the forewing is developed as a free lobe, it projects beneath the humeral angle of the hindwing and thus serves to yoke the two wings together. In the Jugatae group of Lepidoptera, it bears a long finger-like lobe. The jugal region was termed the neala ("new wing") because it is evidently a secondary and recently developed part of the wing.

The auxiliary region containing the axillary sclerites has, in general, the form of a scalene triangle. The base of the triangle (a-b) is the hinge of the wing with the body; the apex (c) is the distal end of the third axillary sclerite; the longer side is anterior to the apex. The point d on the anterior side of the triangle marks the articulation of the radial vein with the second axillary sclerite. The line between d and c is the plica basalis (bf), or fold of the wing at the base of the mediocubital field.

At the posterior angle of the wing base in some Diptera there is a pair of membranous lobes (squamae, or calypteres) known as the alula. The alula is well developed in the house fly. The outer squama (c) arises from the wing base behind the third axillary sclerite (3Ax) and evidently represents the jugal lobe of other insects (A, D); the larger inner squama (d) arises from the posterior scutellar margin of the tergum of the wing-bearing segment and forms a protective, hood-like canopy over the halter. In the flexed wing, the outer squama of the alula is turned upside down above the inner squama, the latter not being affected by the movement of the wing. In many Diptera, a deep incision of the anal area of the wing membrane behind the single vannal vein sets off a proximal alar lobe distal to the outer squama of the alula.

Joints

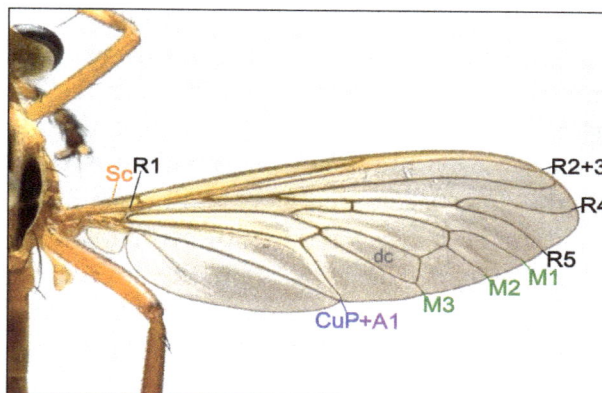

The various movements of the wings, especially in insects that flex their wings horizontally over their backs when at rest, demand a more complicated articular structure at the wing base than a mere hinge of the wing with the body. Each wing is attached to the body by a membranous basal area, but the articular membrane contains a number of small articular sclerites, collectively known as the pteralia. The pteralia include an anterior humeral plate at the base of the costal vein, a group of axillaries (Ax) associated with the subcostal, radial, and vannal veins, and two less definite median plates (m, m') at the base of the mediocubital area. The axillaries are specifically developed only in the wing-flexing insects, where they constitute the flexor mechanism of the wing operated by the flexor muscle arising on the pleuron. Characteristic of the wing base is also a small lobe on the anterior margin of the articular area proximal to the humeral plate, which, in the forewing of some insects, is developed into a large, flat, scale-like flap, the tegula, overlapping the base of the wing. Posteriorly, the articular membrane often forms an ample lobe between the wing and the body, and its margin is generally thickened and corrugated, giving the appearance of a ligament, the so-called axillary cord, continuous mesally with the posterior marginal scutellar fold of the tergal plate bearing the wing.

The articular sclerites, or pteralia, of the wing base of the wing-flexing insects and their relations to the body and the wing veins, shown diagrammatically, are as follows:

- Humeral plates,
- First Axillary,
- Second Axillary,
- Third Axillary,
- Fourth Axillary,
- Median plates (m, m').

The humeral plate is usually a small sclerite on the anterior margin of the wing base, movable and articulated with the base of the costal vein. Odonata have their humeral plates greatly enlarged, with two muscles arising from the episternum inserted into the humeral plates and two from the edge of the epimeron inserted into the axillary plate.

The first axillary sclerite (lAx) is the anterior hinge plate of the wing base. Its anterior part is supported on the anterior notal wing process of the tergum (ANP); its posterior part articulates with the tergal margin. The anterior end of the sclerite is generally produced as a slender arm, the apex of which (e) is always associated with the base of the subcostal vein (Sc), though it is not united with the latter. The body of the sclerite articulates laterally with the second axillary. The second axillary sclerite (2Ax) is more variable in form than the first axillary, but its mechanical relations are no less definite. It is obliquely hinged to the outer margin of the body of the first axillary, and the radial vein (R) is always flexibly attached to its anterior end (d). The second axillary presents both a dorsal and a ventral sclerotization in the wing base; its ventral surface rests upon the fulcral wing process of the pleuron. The second axillary, therefore, is the pivotal sclerite of the wing base, and it specifically manipulates the radial vein.

The third axillary sclerite (3Ax) lies in the posterior part of the articular region of the wing. Its form is highly variable and often irregular, but the third axillary is the sclerite on which is inserted the flexor muscle of the wing (D). Mesally, it articulates anteriorly (f) with the posterior end of the second axillary, and posteriorly (b) with the posterior wing process of the tergum (PNP), or with a small fourth axillary when the latter is present. Distally, the third axillary is prolonged in a process always associated with the bases of the group of veins in the anal region of the wing, here termed the vannal veins (V). The third axillary, therefore, is usually the posterior hinge plate of the wing base and is the active sclerite of the flexor mechanism, which directly manipulates the vannal veins. The contraction of the flexor muscle (D) revolves the third axillary on its mesal articulations (b, f), and thereby lifts its distal arm; this movement produces the flexion of the wing. The fourth axillary sclerite is not a constant element of the wing base. When present, it is usually a small plate intervening between the third axillary and the posterior notal wing process, and is probably a detached piece of the latter.

The median plates (m, m') are also sclerites that are not so definitely differentiated as specific plates as are the three principal axillaries, but they are important elements of the flexor apparatus. They lie in the median area of the wing base distal to the second and third axillaries, and are separated from each other by an oblique line (bf), which forms a prominent convex fold during flexion of the wing. The proximal plate (m) is usually attached to the distal arm of the third axillary and perhaps should be regarded as a part of the latter. The distal plate (m') is less constantly present as a distinct sclerite, and may be represented by a general sclerotization of the base of the mediocubital field of the wing. When the veins of this region are distinct at their bases, they are associated with the outer median plate.

Coupling, Folding and other Features

In many insect species, the forewing and hindwing are coupled together, which improves the aerodynamic efficiency of flight. The most common coupling mechanism (e.g., Hymenoptera and Trichoptera) is a row of small hooks on the forward margin of the hindwing, or "hamuli", which lock onto the forewing, keeping them held together (hamulate coupling). In some other insect species (e.g., Mecoptera, Lepidoptera, and some Trichoptera) the jugal lobe of the forewing covers a portion of the hindwing (jugal coupling), or the margins of the forewing and hindwing overlap broadly (amplexiform coupling), or the hindwing bristles, or frenulum, hook under the retaining structure or retinalucum on the forewing.

When at rest, the wings are held over the back in most insects, which may involve longitudinal folding of the wing membrane and sometimes also transverse folding. Folding may sometimes occur along the flexion lines. Though fold lines may be transverse, as in the hindwings of beetles and earwigs, they are normally radial to the base of the wing, allowing adjacent sections of a wing to be folded over or under each other. The commonest fold line is the jugal fold, situated just behind the third anal vein, although, most Neoptera have a jugal fold just behind vein 3A on the forewings. It is sometimes also present on the hindwings. Where the anal area of the hindwing is large, as in Orthoptera and Blattodea, the whole of this part may be folded under the anterior part of the wing along a vannal fold a little posterior to the claval furrow. In addition, in Orthoptera and Blattodea, the anal area is folded like a fan along the veins, the anal veins being convex, at the crests of the folds, and the accessory veins concave. Whereas the claval furrow and jugal fold are probably homologous in different species, the vannal fold varies in position in different taxa. Folding is produced by a muscle arising on the pleuron and inserted into the third axillary sclerite in such a waythat, when it contracts, the sclerite pivots about its points of articulation with the posterior notal process and the second axillary sclerite.

As a result, the distal arm of the third axillary sclerite rotates upwards and inwards, so that finally its position is completely reversed. The anal veins are articulated with this sclerite in such a way that when it moves they are carried with it and become flexed over the back of the insect. Activity of the same muscle in flight affects the power output of the wing and so it is also important in flight control. In orthopteroid insects, the elasticity of the cuticle causes the vannal area of the wing to fold along the veins. Consequently, energy is expended in unfolding this region when the wings are moved to the flight position. In general, wing extension probably results from the contraction of muscles attached to the basalar sclerite or, in some insects, to the subalar sclerite.

Legs

The typical and usual segments of the insect leg are divided into the coxa, one trochanter, the femur, the tibia, the tarsus, and the pretarsus. The coxa in its more symmetrical form, has the shape of a short cylinder or truncate cone, though commonly it is ovate and may be almost spherical. The proximal end of the coxa is girdled by a submarginal basicostal suture that forms internally a ridge, or basicosta, and sets off a marginal flange, the coxomarginale, or basicoxite. The basicosta strengthens the base of the coxa and is commonly enlarged on the outer wall to give insertion to muscles; on the mesal half of the coxa, however, it is usually weak and often confluent with the coxal margin. The trochanteral muscles that take their origin in the coxa are always attached distal to the basicosta. The coxa is attached to the body by an articular membrane, the coxal corium, which surrounds its base. These two articulations are perhaps the primary dorsal and ventral articular points of the subcoxo-coxal hinge. In addition, the insect coxa has often an anterior articulation with the anterior, ventral end of the trochantin, but the trochantinal articulation does not coexist with a sternal articulation. The pleural articular surface of the coxa is borne on a mesal inflection of the coxal wall. If the coxa is movable on the pleural articulation alone, the coxal articular surface is usually inflected to a sufficient depth to give a leverage to the abductor muscles inserted on the outer rim of the coxal base. Distally the coxa bears an anterior and a posterior articulation with the trochanter. The outer wall of the coxa is often marked by a suture extending from the base to the anterior trochanteral articulation. In some insects the coxal suture falls in line with the pleural suture, and in such cases the coxa appears to be divided into two parts corresponding to the episternum and epimeron of the pleuron. The coxal suture is absent in many insects.

The inflection of the coxal wall bearing the pleural articular surface divides the lateral wall of the basicoxite into a prearticular part and a postarticular part, and the two areas often appear as two marginal lobes on the base of the coxa. The posterior lobe is usually the larger and is termed the meron. The meron may be greatly enlarged by an extension distally in the posterior wall of the coxa; in the Neuroptera, Mecoptera, Trichoptera, and Lepidoptera, the meron is so large that the coxa appears to be divided into an anterior piece, the so-called "coxa genuina," and the meron, but the meron never includes the region of the posterior trochanteral articulation, and the groove delimiting it is always a part of the basicostal suture. A coxa with an enlarged meron has an appearance similar to one divided by a coxal suture falling in line with the pleural suture, but the two conditions are fundamentally quite different and should not be confused. The meron reaches the extreme of its departure from the usual condition in the Diptera. In some of the more generalized flies, as in the Tipulidae, the meron of the middle leg appears as a large lobe of the coxa projecting upward and posteriorly from the coxal base; in higher members of the order it becomes completely separated from the coxa and forms a plate of the lateral wall of the mesothorax.

The trochanter is the basal segment of the telopodite; it is always a small segment in the insect leg, freely movable by a horizontal hinge on the coxa, but more or less fixed to the base of the femur. When movable on the femur the trochantero femoral hinge is usually vertical or oblique in a vertical plane, giving a slight movement of production and reduction at the joint, though only a reductor muscle is present. In the Odonata, both nymphs and adults, there are two trochanteral segments, but they are not movable on each other; the second contains the reductor muscle of the femur. The usual single trochanteral segment of insects, therefore, probably represents the two trochanters of other arthropods fused into one apparent segment, since it is not likely that the primary coxotrochanteral hinge has been lost from the leg. In some of the Hymenoptera a basal subdivision of the femur simulates a second trochanter, but the insertion of the reductor muscle on its base attests that it belongs to the femoral segment, since as shown in the odonate leg, the reductor has its origin in the true second trochanter.

The femur is the third segment of the insect leg, is usually the longest and strongest part of the limb, but it varies in size from the huge hind femur of leaping Orthoptera to a very small segment such as is present in many larval forms. The volume of the femur is generally correlated with the size of the tibial muscles contained within it, but it is sometimes enlarged and modified in shape for other purposes than that of accommodating the tibial muscles. The tibia is characteristically a slender segment in adult insects, only a little shorter than the femur or the combined femur and trochanter. Its proximal end forms a more or less distinct head bent toward the femur, a device allowing the tibia to be flexed close against the under surface of the femur.

The terms profemur, mesofemur and metafemur refer to the femora of the front, middle and hind legs of an insect, respectively. Similarly protibia, mesotibia and metatibia refer to the tibiae of the front, middle and hind legs.

The tarsus of insects corresponds to the penultimate segment of a generalized arthropod limb, which is the segment called the propodite in Crustacea. adult insects it is commonly subdivided into from two to five subsegments, or tarsomeres, but in the Protura, some Collembola, and most holometabolous insect larvae it preserves the primitive form of a simple segment. The subsegments of the adult insect tarsus are usually freely movable on one another by inflected connecting membranes, but the tarsus never has intrinsic muscles. The tarsus of adult pterygote insects having fewer than five subsegments is probably specialized by the loss of one or more subsegments or by a fusion of adjoining subsegments. In the tarsi of Acrididae the long basal piece is evidently composed of three united tarsomeres, leaving the fourth and the fifth. The basal tarsomere is sometimes conspicuously enlarged and is distinguished as the basitarsus. On the under surfaces of the tarsal subsegments in certain Orthoptera there are small pads, the tarsal pulvilli, or euplantulae. The tarsus is occasionally fused with the tibia in larval insects, forming a tibiotarsal segment; in some cases it appears to be eliminated or reduced to a rudiment between the tibia and the pretarsus.

For the most part the femur and tibia are the longest leg segments but variations in the lengths and robustness of each segment relate to their functions. For example, gressorial and cursorial, or walking and running type insects respectively, usually have well-developed femora and tibiae on all legs, whereas jumping (saltatorial) insects such as grasshoppers have disproportionately developed metafemora and metatibiae. In aquatic beetles (Coleoptera) and bugs (Hemiptera), the tibiae and/or tarsi of one or more pairs of legs usually are modified for swimming (natatorial) with fringes of long, slender hairs. Many ground-dwelling insects, such as mole crickets (Orthoptera: Gryllotalpidae), nymphal cicadas (Hemiptera: Cicadidae), and scarab beetles (Scarabaeidae), have the tibiae of the forelegs (protibiae) enlarged and modified for digging (fossorial), whereas the forelegs of some predatory insects, such as mantispid lacewings (Neuroptera) and mantids (Mantodea), are specialized for seizing prey, or raptorial. The tibia and basal tarsomere of each hindleg of honey bees are modified for the collection and carriage of pollen.

Abdomen

The ground plan of the abdomen of an adult insect typically consists of 11–12 segments and is less strongly sclerotized than the head or thorax. Each segment of the abdomen is represented by a sclerotized tergum, sternum, and perhaps a pleurite. Terga are separated from each other and from the adjacent sterna or pleura by a membrane. Spiracles are located in the pleural area. Variation of this ground plan includes the fusion of terga or terga and sterna to form continuous dorsal or ventral shields or a conical tube. Some insects bear a sclerite in the pleural area called a laterotergite. Ventral sclerites are sometimes called laterosternites. During the embryonic stage of many insects and the postembryonic stage of primitive insects, 11 abdominal segments are present. In modern insects there is a tendency toward reduction in the number of the abdominal segments, but the primitive number of 11 is maintained during embryogenesis.Variation in abdominal segment number is considerable. If the Apterygota are considered to be indicative of the ground plan for pterygotes, confusion reigns: adult Protura have 12 segments, Collembola have 6. The orthopteran family Acrididae has 11 segments, and a fossil specimen of Zoraptera has a 10-segmented abdomen.

Generally, the first seven abdominal segments of adults (the pregenital segments) are similar in structure and lack appendages. However, apterygotes (bristletails and silverfish) and many immature aquatic insects have abdominal appendages. Apterygotes possess a pair of styles; rudimentary appendages that are serially homologous with the distal part of the thoracic legs. And, mesally, one or two pairs of protrusible (or exsertile) vesicles on at least some abdominal segments. These vesicles are derived from the coxal and trochanteral endites (inner annulated lobes) of the ancestral abdominal appendages. Aquatic larvae and nymphs may have gills laterally on some to most abdominal segments. Of the rest of the abdominal segments consist of the reproductive and anal parts.

The anal-genital part of the abdomen, known as the terminalia, consists generally of segments 8 or 9 to the abdominal apex. Segments 8 and 9 bear the genitalia; segment 10 is visible as a complete segment in many "lower" insects but always lacks appendages; and the small segment 11 is represented by a dorsal epiproct and pair of ventral paraprocts derived from the sternum. A pair of appendages, the cerci, articulates laterally on segment 11; typically these are annulated and filamentous but have been modified (e.g. the forceps of earwigs) or reduced in different insect orders. An annulated caudal filament, the median appendix dorsalis, arises from the tip of the epiproct in apterygotes, most mayflies (Ephemeroptera), and a few fossil insects. A similar structure in nymphal stoneflies (Plecoptera) is of uncertain homology. These terminal abdominal segments have excretory and sensory functions in all insects, but in adults there is an additional reproductive function.

External Genitalia

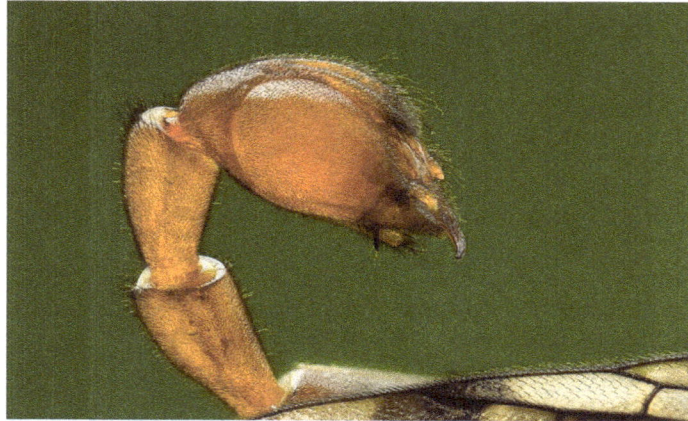

The abdominal terminus of male scorpionflies is enlarged into a "genital bulb".

The organs concerned specifically with mating and the deposition of eggs are known collectively as the external genitalia, although they may be largely internal. The components of the external genitalia of insects are very diverse in form and often have considerable taxonomic value, particularly among species that appear structurally similar in other respects. The male external genitalia have been used widely to aid in distinguishing species, whereas the female external genitalia may be simpler and less varied.

The terminalia of adult female insects include internal structures for receiving the male copulatory organ and his spermatozoa and external structures used for oviposition Most female insects have an egg-laying tube, or ovipositor; it is absent in termites, parasitic lice, many Plecoptera, and most Ephemeroptera. Ovipositors take two forms:

1. True, or appendicular, formed from appendages of abdominal segments 8 and 9;

2. Substitutional, composed of extensible posterior abdominal segments.

Internal

Digestive System

An insect uses its digestive system to extract nutrients and other substances from the food it consumes. Most of this food is ingested in the form of macromolecules and other complex substances like proteins, polysaccharides, fats, and nucleic acids. These macromolecules must be broken down by catabolic reactions into smaller molecules like amino acids and simple sugars before being used by cells of the body for energy, growth, or reproduction. This break-down process is known as digestion. The main structure of an insect's digestive system is a long enclosed tube called the alimentary canal, which runs lengthwise through the body. The alimentary canal directs food in one direction: from the mouth to the anus. It has three sections, each of which performs a different process of digestion. In addition to the alimentary canal, insects also have paired salivary glands and salivary reservoirs. These structures usually reside in the thorax, adjacent to the foregut. The gut is where almost all of insects' digestion takes place. It can be divided into the foregut, midgut and hindgut.

Foregut

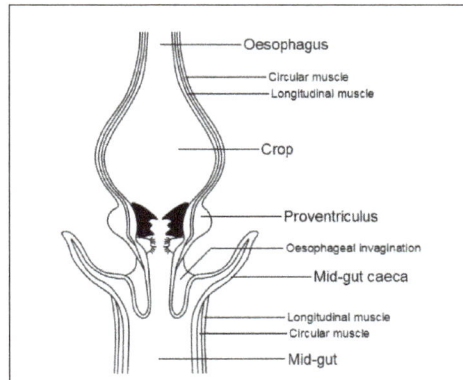

Stylized diagram of insect digestive tract showing malpighian tubule, from an insect of the order Orthoptera.

The first section of the alimentary canal is the foregut, or stomodaeum. The foregut is lined with a cuticular lining made of chitin and proteins as protection from tough food. The foregut includes the buccal cavity (mouth), pharynx, esophagus, and Crop and proventriculus (any part may be highly modified), which both store food and signify when to continue passing onward to the midgut. Here, digestion starts as partially chewed food is broken down by saliva from the salivary glands. As the salivary glands produce fluid and carbohydrate-digesting enzymes (mostly amylases), strong muscles in the pharynx pump fluid into the buccal cavity, lubricating the food like the salivarium does, and helping blood feeders, and xylem and phloem feeders.

From there, the pharynx passes food to the esophagus, which could be just a simple tube passing it on to the crop and proventriculus, and then on ward to the midgut, as in most insects. Alternately, the foregut may expand into a very enlarged crop and proventriculus, or the crop could just be a diverticulum, or fluid filled structure, as in some Diptera species.

Bumblebee defecating.

The salivary glands (element 30 in numbered diagram) in an insect's mouth produce saliva. The salivary ducts lead from the glands to the reservoirs and then forward through the head to an opening called the salivarium, located behind the hypopharynx. By moving its mouthparts (element 32 in numbered diagram) the insect can mix its food with saliva. The mixture of saliva and food then travels through the salivary tubes into the mouth, where it begins to break down. Some insects, like flies, have extra-oral digestion. Insects using extra-oral digestion expel digestive enzymes onto their food to break it down. This strategy allows insects to extract a significant proportion of the available nutrients from the food source.

Midgut

Once food leaves the crop, it passes to the midgut (element 13 in numbered diagram), also known as the mesenteron, where the majority of digestion takes place. Microscopic projections from the midgut wall, called microvilli, increase the surface area of the wall and allow more nutrients to be absorbed; they tend to be close to the origin of the midgut. In some insects, the role of the microvilli and where they are located may vary. For example, specialized microvilli producing digestive enzymes may more likely be near the end of the midgut, and absorption near the origin or beginning of the midgut.

Hindgut

In the hindgut (element 16 in numbered diagram), or proctodaeum, undigested food particles are joined by uric acid to form fecal pellets. The rectum absorbs 90% of the water in these fecal pellets, and the dry pellet is then eliminated through the anus (element 17), completing the process of digestion. The uric acid is formed using hemolymph waste products diffused from the Malpighian tubules (element 20). It is then emptied directly into the alimentary canal, at the junction between the midgut and hindgut. The number of Malpighian tubules possessed by a given insect varies between species, ranging from only two tubules in some insects to over 100 tubules in others.

Respiratory Systems

Insect respiration is accomplished without lungs. Instead, the insect respiratory system uses a system of internal tubes and sacs through which gases either diffuse or are actively pumped, delivering oxygen directly to tissues that need it via their trachea (element 8 in numbered diagram). Since oxygen is delivered directly, the circulatory system is not used to carry oxygen, and is therefore greatly reduced. The insect circulatory system has no veins or arteries, and instead consists of little more than a single, perforated dorsal tube that pulses peristaltically. Toward the thorax, the dorsal tube (element 14) divides into chambers and acts like the insect's heart. The opposite end of the dorsal tube is like the aorta of the insect circulating the hemolymph, arthropods' fluid analog of blood, inside the body cavity. Air is taken in through openings on the sides of the abdomen called spiracles.

There are many different patterns of gas exchange demonstrated by different groups of insects. Gas exchange patterns in insects can range from continuous and diffusive ventilation, to discontinuous gas exchange. During continuous gas exchange, oxygen is taken in and carbon dioxide is released in a continuous cycle. In discontinuous gas exchange, however, the insect takes in oxygen while it is active and small amounts of carbon dioxide are released when the insect is at rest. Diffusive ventilation is simply a form of continuous gas exchange that occurs by diffusion rather than physically taking in the oxygen. Some species of insect that are submerged also have adaptations to aid in respiration. As larvae, many insects have gills that can extract oxygen dissolved in water, while others need to rise to the water surface to replenish air supplies, which may be held or trapped in special structures.

Circulatory System

Insect blood or haemolymph's main function is that of transport and it bathes the insect's body organs. Making up usually less than 25% of an insect's body weight, it transports hormones, nutrients and wastes and has a role in, osmoregulation, temperature control, immunity, storage (water, carbohydrates and fats) and skeletal function. It also plays an essential part in the moulting process. An additional role of the haemolymph in some orders, can be that of predatory defence. It can contain unpalatable and malodourous chemicals that will act as a deterrent to predators. Haemolymph contains molecules, ions and cells; regulating chemical exchanges between tissues, haemolymph is encased in the insect body cavity or haemocoel. It is transported around the body by combined heart (posterior) and aorta (anterior) pulsations, which are located dorsally just under the surface of the body. It differs from vertebrate blood in that it doesn't contain any red blood cells and therefore is without high oxygen carrying capacity, and is more similar to lymph found in vertebrates.

Body fluids enter through one-way valved ostia, which are openings situated along the length of the combined aorta and heart organ. Pumping of the haemolymph occurs by waves of peristaltic contraction, originating at the body's posterior end, pumping forwards into the dorsal vessel, out via the aorta and then into the head where it flows out into the haemocoel. The haemolymph is circulated to the appendages unidirectionally with the aid of muscular pumps or accessory pulsatile organs usually found at the base of the antennae or wings and sometimes in the legs, with pumping rates accelerating with periods of increased activity. Movement of haemolymph is particularly important for thermoregulation in orders such as Odonata, Lepidoptera, Hymenoptera and Diptera.

Endocrine System

These glands are part of the endocrine system:

1. Neurosecretory cells,

2. Corpora cardiaca,

3. Prothoracic glands,

4. Corpora allata.

Reproductive System

Female

Female insects are able make eggs, receive and store sperm, manipulate sperm from different males, and lay eggs. Their reproductive systems are made up of a pair of ovaries, accessory glands, one or more spermathecae, and ducts connecting these parts. The ovaries make eggs and accessory glands produce the substances to help package and lay the eggs. Spermathecae store sperm for varying periods of time and, along with portions of the oviducts, can control sperm use. The ducts and spermathecae are lined with a cuticle.

The ovaries are made up of a number of egg tubes, called ovarioles, which vary in size and number by species. The number of eggs that the insect is able to make vary by the number of ovarioles with the rate that eggs can be developed being also influenced by ovariole design. In meroistic ovaries, the eggs-to-be divide repeatedly and most of the daughter cells become helper cells for a single oocyte in the cluster. In panoistic ovaries, each egg-to-be produced by stem germ cells develops into an oocyte; there are no helper cells from the germ line. Production of eggs by panoistic ovaries tends to be slower than that by meroistic ovaries.

Accessory glands or glandular parts of the oviducts produce a variety of substances for sperm maintenance, transport, and fertilization, as well as for protection of eggs. They can produce glue and protective substances for coating eggs or tough coverings for a batch of eggs called oothecae. Spermathecae are tubes or sacs in which sperm can be stored between the time of mating and the time an egg is fertilized. Paternity testing of insects has revealed that some, and probably many, female insects use the spermatheca and various ducts to control or bias sperm used in favor of some males over others.

Male

The main component of the male reproductive system is the testis, suspended in the body cavity by tracheae and the fat body. The more primitive apterygote insects have a single testis, and in some lepidopterans the two maturing testes are secondarily fused into one structure during the later stages of larval development, although the ducts leading from them remain separate. However, most male insects have a pair of testes, inside of which are sperm tubes or follicles that are enclosed within a membranous sac. The follicles connect to the vas deferens by the vas efferens, and the two tubular vasa deferentia connect to a median ejaculatory duct that leads to the outside. A portion of the vas deferens is often enlarged to form the seminal vesicle, which stores the sperm before they are discharged into the female. The seminal vesicles have glandular linings that secrete nutrients for nourishment and maintenance of the sperm. The ejaculatory duct is derived from an invagination of the epidermal cells during development and, as a result, has a cuticular lining. The terminal portion of the ejaculatory duct may be sclerotized to form the intromittent organ, the aedeagus. The remainder of the male reproductive system is derived from embryonic mesoderm, except for the germ cells, or spermatogonia, which descend from the primordial pole cells very early during embryogenesis. The aedeagus can be quite pronounced or *de minimis*. The base of the aedeagus may be the partially sclerotized phallotheca, also called the phallosoma or theca. In some species the phallotheca contains a space, called the endosoma (internal holding pouch), into which the tip end of the aedeagus may be withdrawn (retracted). The vas deferens is sometimes drawn into (folded into) the phallotheca together with a seminal vesicle.

Internal Morphology of Different Taxa

Blattodea

Cockroaches are most common in tropical and subtropical climates. Some species are in close association with human dwellings and widely found around garbage or in the kitchen. Cockroaches are generally omnivorous with the exception of the wood-eating species such as *Cryptocercus*; these roaches are incapable of digesting cellulose themselves, but have symbiotic relationships with various protozoans and bacteria that digest the cellulose, allowing them to extract the nutrients. The similarity of these symbionts in the genus *Cryptocercus* to those in termites are such that it has been suggested that they are more closely related to termites than to other cockroaches, and current research strongly supports this hypothesis of relationships. All species studied so far carry the obligate mutualistic endosymbiont bacterium *Blattabacterium*, with the exception of *Nocticola australiensis*, an Australian cave dwelling species without eyes, pigment or wings, and which recent genetic studies indicates are very primitive cockroaches.

Cockroaches, like all insects, breathe through a system of tubes called *tracheae*. The tracheae of insects are attached to the spiracles, excluding the head. Thus cockroaches, like all insects, are not dependent on the mouth and windpipe to breathe. The valves open when the CO_2 level in the insect rises to a high level; then the CO_2 diffuses out of the tracheae to the outside and fresh O_2 diffuses in. Unlike in vertebrates that depend on blood for transporting O_2 and CO_2, the tracheal system brings the air directly to cells, the tracheal tubes branching continually like a tree until their finest divisions, tracheoles, are associated with each cell, allowing gaseous oxygen to dissolve in the cytoplasm lying across the fine cuticle lining of the tracheole. CO_2 diffuses out of the cell into the tracheole. While cockroaches do not have lungs and thus do not actively breathe in the vertebrate lung manner, in some very large species the body musculature may contract rhythmically to forcibly move air out and in the spiracles; this may be considered a form of breathing.

Coleoptera

The digestive system of beetles is primarily based on plants, which they for the most part feed upon, with mostly the anterior midgut performing digestion. However, in predatory species (e.g., Carabidae) most digestion occurs in the crop by means of midgut enzymes. In Elateridae species, the predatory larvae defecate enzymes on their prey, with digestion being extraorally. The alimentary canal basically comprises a short narrow pharynx, a widened expansion, the crop and a poorly developed gizzard. After there is a midgut, that varies in dimensions between species, with a large amount of cecum, with a hingut, with varying lengths. There are typically four to six Malpighian tubules.

The nervous system in beetles contains all the types found in insects, varying between different species. With three thoracic and seven or eight abdominal ganglia can be distinguished to that in which all the thoracic and abdominal ganglia are fused to form a composite structure. Oxygen is obtained via a tracheal system. Air enters a series of tubes along the body through openings called spiracles, and is then taken into increasingly finer fibers. Pumping movements of the body force the air through the system. Some species of diving beetles (Dytiscidae) carry a bubble of air with them whenever they dive beneath the water surface. This bubble may be held under the elytra or it may be trapped against the body using specialized hairs. The bubble usually covers one or more spiracles so the insect can breathe air from the bubble while submerged. An air bubble provides an insect with only a short-term supply of oxygen, but thanks to its unique physical properties, oxygen will diffuse into the bubble and displacing the nitrogen, called passive diffusion, however the volume of the bubble eventually diminishes and the beetle will have to return to the surface.

Like other insect species, beetles have hemolymph instead of blood. The open circulatory system of the beetle is driven by a tube-like heart attached to the top inside of the thorax.

Different glands specialize for different pheromones produced for finding mates. Pheromones from species of Rutelinea are produced from epithelial cells lining the inner surface of the apical abdominal segments or amino acid based pheromones of Melolonthinae from eversible glands on the abdominal apex. Other species produce different types of pheromones. Dermestids produce esters, and species of Elateridae produce fatty-acid-derived aldehydes and acetates. For means of finding a mate also, fireflies (Lampyridae) utilized modified fat body cells with transparent surfaces backed with reflective uric acid crystals to biosynthetically produce light, or bioluminescence. The light produce is highly efficient, as it is produced by oxidation of luciferin by the enzymes luciferase in the presence of ATP (adenosine triphospate) and oxygen, producing oxyluciferin, carbon dioxide, and light.

A notable number of species have developed special glands that produce chemicals for deterring predators. The Ground beetle's (of Carabidae) defensive glands, located at the posterior, produce a variety of hydrocarbons, aldehydes, phenols, quinones, esters, and acids released from an opening at the end of the abdomen. While African carabid beetles (e.g., *Anthia* some of which used to comprise the genus *Thermophilum*) employ the same chemicals as ants: formic acid. While Bombardier beetles have well developed, like other carabid beetles, pygidial glands that empty from the lateral edges of the intersegment membranes between the seventh and eighth abdominal segments. The gland is made of two containing chambers. The first holds hydroquinones and hydrogen peroxide, with the second holding just hydrogen peroxide plus catalases. These chemicals mix and result in an explosive ejection, forming temperatures of around 100 C, with the breakdown of hydroquinone to $H_2 + O_2$ + quinone, with the O_2 propelling the excretion.

Tympanal organs are hearing organs. Such an organ is generally a membrane (tympanum) stretched across a frame backed by an air sac and associated sensory neurons. In the order Coleoptera, tympanal organs have been described in at least two families. Several species of the genus Cicindela in the family Cicindelidae have ears on the dorsal surface of the first abdominal segment beneath the wing; two tribes in the family Dynastinae (Scarabaeidae) have ears just beneath the pronotal shield or neck membrane. The ears of both families are to ultrasonic frequencies, with strong evidence that they function to detect the presence of bats via their ultrasonic echolocation. Even though beetles constitute a large order and live in a variety of niches, examples of hearing is surprisingly lacking in species, though it is likely that most are just undiscovered.

Dermaptera

The neuroendocrine system is typical of insects. There is a brain, a subesophageal ganglion, three thoracic ganglia, and six abdominal ganglia. Strong neuron connections connect the neurohemal corpora cardiaca to the brain and frontal ganglion, where the closely related median corpus allatum produces juvenile hormone III in close proximity to the neurohemal dorsal aorta. The digestive system of earwigs is like all other insects, consisting of a fore-, mid-, and hindgut, but earwigs lack gastric caecae which are specialized for digestion in many species of insect. Long, slender (extratory) malpighian tubules can be found between the junction of the mid- and hind gut.

The reproductive system of females consist of paired ovaries, lateral oviducts, spermatheca, and a genital chamber. The lateral ducts are where the eggs leave the body, while the spermatheca is where sperm is stored. Unlike other insects, the gonopore, or genital opening is behind the seventh abdominal segment. The ovaries are primitive in that they are polytrophic (the nurse cells and oocytes alternate along the length of the ovariole). In some species these long ovarioles branch off the lateral duct, while in others, short ovarioles appear around the duct.

Diptera

The genitalia of female flies are rotated to a varying degree from the position found in other insects. In some flies this is a temporary rotation during mating, but in others it is a permanent torsion of the organs that occurs during the pupal stage. This torsion may lead to the anus being located below the genitals, or, in the case of 360° torsion, to the sperm duct being wrapped around the gut, despite the external organs being in their usual position. When flies mate, the male initially flies on top of the female, facing in the same direction, but then turns round to face in the opposite direction. This forces the male to lie on its back in order for its genitalia to remain engaged with those of the female, or the torsion of the male genitals allows the male to mate while remaining upright. This leads to flies having more reproduction abilities than most insects and at a much quicker rate. Flies come in great populations due to their ability to mate effectively and in a short period of time especially during the mating season.

The female lays her eggs as close to the food source as possible, and development is very rapid, allowing the larva to consume as much food as possible in a short period of time before transforming into the adult. The eggs hatch immediately after being laid, or the flies are ovoviviparous, with the larva hatching inside the mother. Larval flies, or maggots, have no true legs, and little demarcation between the thorax and abdomen; in the more derived species, the head is not clearly distinguishable from the rest of the body. Maggots are limbless, or else have small prolegs. The eyes and antennae are reduced or absent, and the abdomen also lacks appendages such as cerci. This lack of features is an adaptation to a food-rich environment, such as within rotting organic matter, or as an endoparasite. The pupae take various forms, and in some cases develop inside a silk cocoon. After emerging from the pupa, the adult fly rarely lives more than a few days, and serves mainly to reproduce and to disperse in search of new food sources.

Lepidoptera

In reproductive system of butterflies and moths, the male genitalia are complex and unclear. In females there are three types of genitalia based on the relating taxa: monotrysian, exoporian, and dytresian. In the monotrysian type there is an opening on the fused segments of the sterna 9 and 10, which act as insemination and oviposition. In the exoporian type (in Hepaloidae and Mnesarchaeoidea) there are two separate places for insemination and oviposition, both occurring on the same sterna as the monotrysian type, 9/10. In most species the genitalia are flanked by two soft lobes, although they may be specialized and sclerotized in some species for ovipositing in area such as crevices and inside plant tissue. Hormones and the glands that produce them run the development of butterflies and moths as they go through their life cycle, called the endocrine system. The first insect hormone PTTH (Prothoracicotropic hormone) operates the species life cycle and diapause. This hormone is produced by corpora allata and corpora cardiaca, where it is also stored. Some glands are specialized to perform certain task such as producing silk or producing saliva in the palpi. While the corpora cardiaca produce PTTH, the corpora allata also produces jeuvanile hormones, and the prothorocic glands produce moulting hormones.

In the digestive system, the anterior region of the foregut has been modified to form a pharyngial sucking pump as they need it for the food they eat, which are for the most part liquids. An esophagus follows and leads to the posterior of the pharynx and in some species forms a form of crop. The midgut is short and straight, with the hindgut being longer and coiled. Ancestors of lepidopteran species, stemming from Hymenoptera, had midgut ceca, although this is lost in current butterflies and moths. Instead, all the digestive enzymes other than initial digestion, are immobilized at the surface of the midgut cells. In larvae, long-necked and stalked goblet cells are found in the anterior and posterior midgut regions, respectively. In insects, the goblet cells excrete positive potassium ions, which are absorbed from leaves ingested by the larvae. Most butterflies and moths display the usual digestive cycle, however species that have a different diet require adaptations to meet these new demands.

In the circulatory system, hemolymph, or insect blood, is used to circulate heat in a form of thermoregulation, where muscles contraction produces heat, which is transferred to the rest of the body when conditions are unfavorable. In lepidopteran species, hemolymph is circulated through the veins in the wings by some form of pulsating organ, either by the heart or by the intake of air into the trachea. Air is taken in through spiracles along the sides of the abdomen and thorax supplying the trachea with oxygen as it goes through the lepidopteran's respiratory system. There are three different tracheae supplying oxygen diffusing oxygen throughout the species body: The dorsal, ventral, and visceral. The dorsal tracheae supply oxygen to the dorsal musculature and vessels, while the ventral tracheae supply the ventral musculature and nerve cord, and the visceral tracheae supply the guts, fat bodies, and gonads.

Insect Physiology

Insect physiology includes the physiology and biochemistry of insect organ systems.

Although diverse, insects are quite similar in overall design, internally and externally. The insect is made up of three main body regions (tagmata), the head, thorax and abdomen. The head comprises six fused segments with compound eyes, ocelli, antennae and mouthparts, which differ according to the insect's particular diet, e.g. grinding, sucking, lapping and chewing. The thorax is made up of three segments: the pro, meso and meta thorax, each supporting a pair of legs which may also differ, depending on function, e.g. jumping, digging, swimming and running. Usually the middle and the last segment of the thorax have paired wings. The abdomen generally comprises eleven segments and contains the digestive and reproductive organs. A general overview of the internal structure and physiology of the insect is presented, including digestive, circulatory, respiratory, muscular, endocrine and nervous systems, as well as sensory organs, temperature control, flight and molting.

Digestive System

An insect uses its digestive system to extract nutrients and other substances from the food it consumes. Most of this food is ingested in the form of macromolecules and other complex substances (such as proteins, polysaccharides, fats, and nucleic acids) which must be broken down by catabolic reactions into smaller molecules (i.e. amino acids, simple sugars, etc.) before being used by cells of the body for energy, growth, or reproduction. This break-down process is known as digestion.

The insect's digestive system is a closed system, with one long enclosed coiled tube called the alimentary canal which runs lengthwise through the body. The alimentary canal only allows food to enter the mouth, and then gets processed as it travels toward the anus. The alimentary canal has specific sections for grinding and food storage, enzyme production, and nutrient absorption. Sphincters control the food and fluid movement between three regions. The three regions include the foregut (stomatodeum)(27,) the midgut (mesenteron)(13), and the hindgut (proctodeum)(16).

In addition to the alimentary canal, insects also have paired salivary glands and salivary reservoirs. These structures usually reside in the thorax (adjacent to the fore-gut). The salivary glands (30) produce saliva; the salivary ducts lead from the glands to the reservoirs and then forward through the head to an opening called the salivarium behind the hypopharynx; which movements of the mouthparts help mix saliva with food in the buccal cavity. Saliva mixes with food, which travels through salivary tubes into the mouth, beginning the process of breaking it down.

The stomatedeum and proctodeum are invaginations of the epidermis and are lined with cuticle (intima). The mesenteron is not lined with cuticle but with rapidly dividing and therefore constantly replaced, epithelial cells. The cuticle sheds with every moult along with the exoskeleton. Food is moved down the gut by muscular contractions called peristalsis.

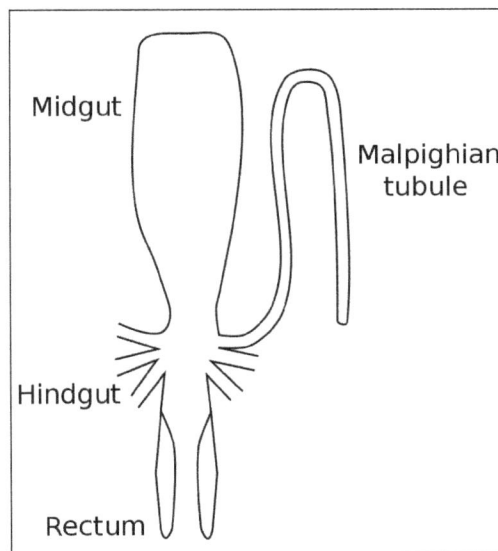

Stylised diagram of insect digestive tract showing Malpighian tubule (Orthopteran type):

1. Stomatodeum (foregut): This region stores, grinds and transports food to the next region. Included in this are the buccal cavity, the pharynx, the oesophagus, the crop (stores food), and proventriculus or gizzard (grinds food). Salivary secretions from the labial glands dilute the ingested food. In mosquitoes (Diptera), which are blood-feeding insects, anticoagulants and blood thinners are also released here.

2. Mesenteron (midgut): Digestive enzymes in this region are produced and secreted into the lumen and here nutrients are absorbed into the insect's body. Food is enveloped by this part of the gut as it arrives from the foregut by the peritrophic membrane which is a mucopolysaccharide layer secreted from the midgut's epithelial cells. It is thought that this membrane prevents food pathogens from contacting the epithelium and attacking

the insects' body. It also acts as a filter allowing small molecules through, but preventing large molecules and particles of food from reaching the midgut cells. After the large substances are broken down into smaller ones, digestion and consequent nutrient absorption takes place at the surface of the epithelium. Microscopic projections from the mid-gut wall, called microvilli, increase surface area and allow for maximum absorption of nutrients.

3. Proctodeum (hindgut): This is divided into three sections; the anterior is the ileum, the middle portion, the colon, and the wider, posterior section is the rectum. This extends from the pyloric valve which is located between the mid and the hindgut to the anus. Here absorption of water, salts and other beneficial substances take place before excretion. Like other animals, the removal of toxic metabolic waste requires water. However, for very small animals like insects, water conservation is a priority. Because of this, blind-ended ducts called Malpighian tubules come into play. These ducts emerge as evaginations at the anterior end of the hindgut and are the main organs of osmoregulation and excretion. These extract the waste products from the haemolymph, in which all the internal organs are bathed). These tubules continually produce the insect's uric acid, which is transported to the hindgut, where important salts and water are re-absorbed by both the hindgut and rectum. Excrement is then voided as insoluble and non-toxic uric acid granules. Excretion and osmoregulation in insects are not orchestrated by the Malpighian tubules alone, but require a joint function of the ileum and/or rectum.

Muscular System

Many insects are able to lift twenty times their own body weight like Rhinoceros beetle and may jump distances that are many times greater than their own length. This is because their energy output is high in relation to their body mass.

The muscular system of insects ranges from a few hundred muscles to a few thousand. Unlike vertebrates that have both smooth and striated muscles, insects have only striated muscles. Muscle cells are amassed into muscle fibers and then into the functional unit, the muscle. Muscles are attached to the body wall, with attachment fibers running through the cuticle and to the epicuticle, where they can move different parts of the body including appendages such as wings. The muscle fiber has many cells with a plasma membrane and outer sheath or sarcolemma. The sarcolemma is invaginated and can make contact with the tracheole carrying oxygen to the muscle fiber. Arranged in sheets or cylindrically, contractile myofibrils run the length of the muscle fiber. Myofibrils comprising a fine actin filament enclosed between a thick pair of myosin filaments slide past each other instigated by nerve impulses.

Muscles can be divided into four categories:

1. Visceral: These muscles surround the tubes and ducts and produce peristalsis as demonstrated in the digestive system.

2. Segmental: Causing telescoping of muscle segments required for moulting, increase in body pressure and locomotion in legless larvae.

3. Appendicular: Originating from either the sternum or the tergum and inserted on the coxae these muscles move appendages as one unit. These are arranged segmentally and usually

in antagonistic pairs. Appendage parts of some insects, e.g. the galea and the lacinia of the maxillae, only have flexor muscles. Extension of these structures is by haemolymph pressure and cuticle elasticity.

4. Flight: Flight muscles are the most specialised category of muscle and are capable of rapid contractions. Nerve impulses are required to initiate muscle contractions and therefore flight. These muscles are also known as neurogenic or synchronous muscles. This is because there is a one-to-one correspondence between action potentials and muscle contractions. In insects with higher wing stroke frequencies the muscles contract more frequently than at the rate that the nerve impulse reaches them and are known as asynchronous muscles.

Flight has allowed the insect to disperse, escape from enemies and environmental harm, and colonise new habitats. One of the insect's key adaptations is flight, the mechanics of which differ from those of other flying animals because their wings are not modified appendages. Fully developed and functional wings occur only in adult insects. To fly, gravity and drag (air resistance to movement) have to be overcome. Most insects fly by beating their wings and to power their flight they have either direct flight muscles attached to the wings, or an indirect system where there is no muscle-to-wing connection and instead they are attached to a highly flexible box-like thorax.

Direct flight muscles generate the upward stroke by the contraction of the muscles attached to the base of the wing inside the pivotal point. Outside the pivotal point the downward stroke is generated through contraction of muscles that extend from the sternum to the wing. Indirect flight muscles are attached to the tergum and sternum. Contraction makes the tergum and base of the wing pull down. In turn this movement lever the outer or main part of the wing in strokes upward. Contraction of the second set of muscles, which run from the back to the front of the thorax, powers the downbeat. This deforms the box and lifts the tergum.

Endocrine System

Hormones are the chemical substances that are transported in the insect's body fluids (haemolymph) that carry messages away from their point of synthesis to sites where physiological processes are influenced. These hormones are produced by glandular, neuroglandular and neuronal centres. Insects have several organs that produce hormones, controlling reproduction, metamorphosis and moulting. It has been suggested that a brain hormone is responsible for caste dermination in termites and diapause interruption in some insects.

Four endocrine centers have been identified:

1. Neurosecretory cells in the brain can produce one or more hormones that affect growth, reproduction, homeostasis and metamorphosis.

2. Corpora cardiaca are a pair of neuroglandular bodies that are found behind the brain and on either sides of the aorta. These not only produce their own neurohormones but they store and release other neurohormones including PTTH prothoracicotropic hormone (brain hormone), which stimulates the secretory activity of the prothoracic glands, playing an integral role in moulting.

3. Prothoracic glands are diffuse, paired glands located at the back of the head or in the thorax. These glands secrete an ecdysteroid called ecdysone, or the moulting hormone, which initiates the epidermal moulting process. Additionally it plays a role in accessory reproductive glands in the female, differentiation of ovarioles and in the process of egg production.

4. Corpora allata are small, paired glandular bodies originating from the epithelium located on either side of the foregut. They secrete the juvenile hormone, which regulate reproduction and metamorphosis.

Nervous System

Insects have a complex nervous system which incorporates a variety of internal physiological information as well as external sensory information. As in the case of vertebrates, the basic component is the neuron or nerve cell. This is made up of a dendrite with two projections that receive stimuli and an axon, which transmits information to another neuron or organ, like a muscle. As for vertebrates, chemicals (neurotransmitters such as acetylcholine and dopamine) are released at synapses.

Central Nervous System

An insect's sensory, motor and physiological processes are controlled by the central nervous system along with the endocrine system. Being the principal division of the nervous system, it consists of a brain, a ventral nerve cord and a subesophageal ganglion which is connected to the brain by two nerves, extending around each side of the oesophagus.

The brain has three lobes:

- Protocerebrum, innervating the compound eyes and the ocelli.

- Deutocerebrum, innervating the antennae.

- Tritocerebrum, innervating the foregut and the labrum.

The ventral nerve cord extends from the suboesophageal ganglion posteriorly. A layer of connective tissue called the neurolemma covers the brain, ganglia, major peripheral nerves and ventral nerve cords.

The head capsule (made up of six fused segments) has six pairs of ganglia. The first three pairs are fused into the brain, while the three following pairs are fused into the subesophageal ganglion. The thoracic segments have one ganglion on each side, which are connected into a pair, one pair per segment. This arrangement is also seen in the abdomen but only in the first eight segments. Many species of insects have reduced numbers of ganglia due to fusion or reduction. Some cockroaches have just six ganglia in the abdomen, whereas the wasp *Vespa crabro* has only two in the thorax and three in the abdomen. And some, like the house fly *Musca domestica*, have all the body ganglia fused into a single large thoracic ganglion. The ganglia of the central nervous system act as the coordinating centres with their own specific autonomy where each may coordinate impulses in specified regions of the insect's body.

Peripheral Nervous System

This consists of motor neuron axons that branch out to the muscles from the ganglia of the central nervous system, parts of the sympathetic nervous system and the sensory neurons of the cuticular sense organs that receive chemical, thermal, mechanical or visual stimuli from the insects environment. The sympathetic nervous system includes nerves and the ganglia that innervate the gut both posteriorly and anteriorly, some endocrine organs, the spiracles of the tracheal system and the reproductive organs.

Sensory organs

Chemical senses include the use of chemoreceptors, related to taste and smell, affecting mating, habitat selection, feeding and parasite-host relationships. Taste is usually located on the mouthparts of the insect but in some insects, such as bees, wasps and ants, taste organs can also be found on the antennae. Taste organs can also be found on the tarsi of moths, butterflies and flies. Olfactory sensilla enable insects to smell and are usually found in the antennae. Chemoreceptor sensitivity related to smell in some substances, is very high and some insects can detect particular odours that are at low concentrations miles from their original source.

Mechanical senses provide the insect with information that may direct orientation, general movement, flight from enemies, reproduction and feeding and are elicited from the sense organs that are sensitive to mechanical stimuli such as pressure, touch and vibration. Hairs (setae) on the cuticle are responsible for this as they are sensitive to vibration touch and sound.

Hearing structures or tympanal organs are located on different body parts such as, wings, abdomen, legs and antennae. These can respond to various frequencies ranging from 100 Hz to 240 kHz depending on insect species. Many of the joints of the insect have tactile setae that register movement. Hair beds and groups of small hair like sensilla, determine proprioreception or information about the position of a limb, and are found on the cuticle at the joints of segments and legs. Pressure on the body wall or strain gauges are detected by the campiniform sensilla and internal stretch receptors sense muscle distension and digestive system stretching.

The compound eye and the ocelli supply insect vision. The compound eye consists of individual light receptive units called ommatidia. Some ants may have only one or two, however dragonflies may have over 10,000. The more ommatidia the greater the visual acuity. These units have a clear lens system and light sensitive retina cells. By day, the image flying insects receive is made up of a mosaic of specks of differing light intensity from all the different ommatidia. At night or dusk, visual acuity is sacrificed for light sensitivity. The ocelli are unable to form focused images but are sensitive mainly, to differences in light intensity. Colour vision occurs in all orders of insects. Generally insects see better at the blue end of the spectrum than at the red end. In some orders sensitivity ranges can include ultraviolet.

A number of insects have temperature and humidity sensors and insects being small, cool more quickly than larger animals. Insects are generally considered cold-blooded or ectothermic, their body temperature rising and falling with the environment. However, flying insects raise their body temperature through the action of flight, above environmental temperatures.

The body temperature of butterflies and grasshoppers in flight may be 5 °C or 10 °C above environmental temperature, however moths and bumblebees, insulated by scales and hair, during flight, may raise flight muscle temperature 20–30 °C above the environment temperature. Most flying insects have to maintain their flight muscles above a certain temperature to gain power enough to fly. Shivering, or vibrating the wing muscles allow larger insects to actively increase the temperature of their flight muscles, enabling flight.

Until very recently, no one had ever documented the presence of nociceptors (the cells that detect and transmit sensations of pain) in insects, though recent findings of nociception in larval fruit flies challenges this and raises the possibility that some insects may be capable of feeling pain.

Life Cycle

An insect's life-cycle can be divided into three types:

- Ametabolous, no metamorphosis, these insects are primitively wingless where the only difference between adult and nymph is size, e.g. order: Thysanura (silverfish).

- Hemimetabolous, or incomplete metamorphosis. The terrestrial young are called nymphs and aquatic young are called naiads. Insect young are usually similar to the adult. Wings appear as buds on the nymphs or early instars. When the last moult is completed the wings expand to the full adult size, e.g. order: Odonata (dragonflies).

- Holometabolus, or complete metamorphosis. These insects have a different form in their immature and adult stages, have different behaviors and live in different habitats. The immature form is called larvae and remains similar in form but increases in size. They usually have chewing mouthparts even if the adult form mouth parts suck. At the last larval instar phase the insect forms into a pupa, it doesn't feed and is inactive, and here wing development is initiated, and the adult emerges e.g. order: Lepidoptera (butterflies and moths).

Moulting

As an insect grows it needs to replace the rigid exoskeleton regularly. Moulting may occur up to three or four times or, in some insects, fifty times or more during its life. A complex process controlled by hormones, it includes the cuticle of the body wall, the cuticular lining of the tracheae, foregut, hindgut and endoskeletal structures.

The stages of molting:

- Apolysis: Moulting hormones are released into the haemolymph and the old cuticle separates from the underlying epidermal cells. The epidermis increases in size due to mitosis and then the new cuticle is produced. Enzymes secreted by the epidermal cells digest the old endocuticle, not affecting the old sclerotised exocuticle.

- Ecdysis: This begins with the splitting of the old cuticle, usually starting in the midline of the thorax's dorsal side. The rupturing force is mostly from haemolymph pressure that has been forced into thorax by abdominal muscle contractions caused by the insect swallowing air or water. After this the insect wriggles out of the old cuticle.

- Sclerotisation: After emergence the new cuticle is soft and this a particularly vulnerable time for the insect as its hard protective coating is missing. After an hour or two the exocuticle hardens and darkens. The wings expand by the force of haemolymph into the wing veins.

Insect Nervous System

The basic insect nervous system bauplan consists of a series of body segments, each equipped with a pair of connected ganglia, with a paired nerve cord connecting adjacent ganglia in each segment. The ganglia are bulbous structures consisting of neuron cell-bodies and supporting or glial cells and acts as a local processor or computer. the ganglia are interconnected by neurons, constituing a computer network. This plan is variously modified in the various types, but in all cases the ganglia of the head segment form a fused mass, situated above the oesophagus (esophagus) of the gut, and called the supraoesophageal ganglion. This is connected by a pair of nerve trunks (connectives or commissures) that course around the oesophagus on either side and join to the suboesophageal ganglion (SOG or SEG) situated beneath the oesophagus. Many consider only the supraoesophageal ganglion to constitute the insect brain, others (including myself) consider the SOG as part of the brain.

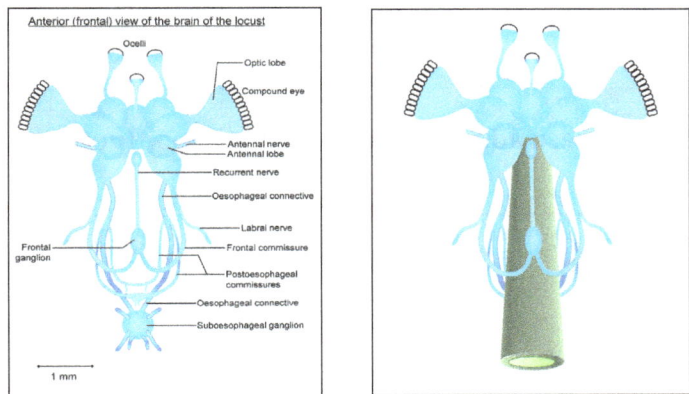

The relative size of the brain species. That of the diving beetle Dytiscus is about 1/400 of the total body size, that of the ant (Formica) is about 1/280, and that of the bee about 1/174. The brain is generally larger in those insects that have more complex social lives Although much smaller than a human brain, containing only one thousandth as many cells, it is still immensely complex. There is also less replication of function - fewer neurons perform each function.

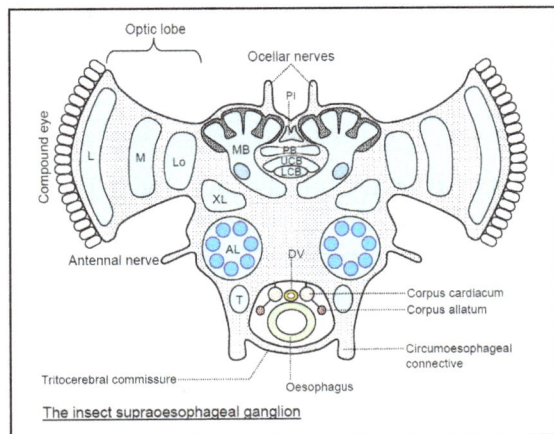

The insect supraoesophageal ganglion

AL, antennal lobe; DV, dorsal blood vessel; L, lamina; LCB, lower central body; Lo, lobula; M, medulla; MB, mushroom body; PB, procerebral bridge; PI, pars intercerebralis; T, tritocerebrum; UCB, upper central body; XL: accessory lobe.

The supraoesophageal ganglion consists of several fused ganglia or lobes. The paired ganglia of the first (frontmost) head segment form the protocerebrum, concerned with vision, time-keeping, higher functions, memory and combining information from different sensory modalities. Those of the segment segment form the deutocerebrum, which is concerned with processing sensory inputs from the antennae, and also the labial palps and parts of the tegument (body wall). the ganglia of the third segment form the relatively small tritocerebrum.

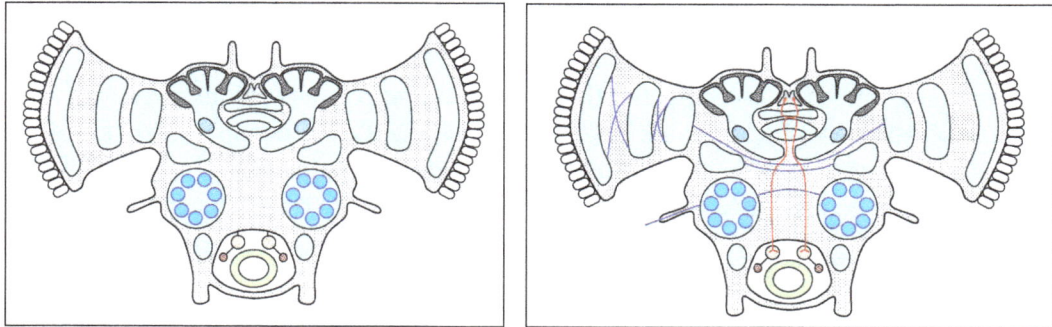

Above: Some fibre tracts have been added: tracts in the optic lobe (notice the crossovers) and connectives between pairs of lobes.

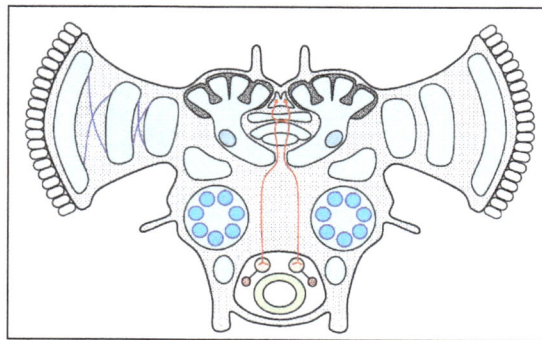

Above: Neurosecretory cells with cell bodies in the pars intercerebralis added. The axons of these neurons terminate in the corpora cardiaca where they releases hormones: part of the insect neuroendocrine system.

Protocerebrum

The optic lobes of the fly (an insect with particularly good vision) contains about 76% of the brain's neurons. The optic lobe connects directly to the sensory cells (retinula cells) in the retina of the compound eye. It contains three distinct regions (neuropils): the lamina, medulla and lobula, where processing of visual signals begins. The protocerebrum also receives inputs via the ocelli, when present, via the ocellar nerves.

The mushroom bodies (MB, corpora pedunculata, 'stalked bodies') are best developed in social insects, making up 20% of the brain of the bee and 50% of the brain of worker ants (Formica). These are thought to function as higher centres responsible for the most sophisticated computations occurring in the insect brain. Each consists of a topmost cap and a stalk or peduncle (which branches into at least two lobes). The cap consists of a pair of cup-like structures, the medial calyx and the lateral calyx (plural of calyx is calyces). The mushroom bodies receive sensory inputs from the lobula of the optic lobe and from the antennal lobes of the deutocerebrum. Most sensory inputs enter the MB through the calyx. There are about 1000 to 100 000 specialised neurons, called Kenyon cells, in each mushroom body. These neurons have tree-like branching dendrites which receive inputs in the calyces of the MB, a single axon which extends down the stalk of the MB and then gives of branches to two lobes of the MB. Dragonfly mushroom bodies have no calyces and no Kenyon cells. The mushroom bodies are also involved in learning, and in the honeybee have been shown to process memories, transferring data from short-term memory (STM) into long-term memory (LTM).

The central body receives inputs from the mushroom bodies and integrates sensory inputs from different sensory modalities (such as small and vision) - so-called multimodal sensory perception. It functions as an activating centre, switching on appropriate locomotor activity patterns which are central programs located in the thoracic ganglia. That is it instructs the thoracic ganglia which programs to run - programs that control the legs and wings. These hard-wired programs are sometimes called central pattern generators and require no sensory input for their execution, though sensory inputs may start and stop these programs or modify them slightly.

The pars intercerebralis is a mass of cell bodies, including neurosecretory cells which send their axons to the pair of corpora cardiaca. The corpora cardiac are sometimes fused into a single medial ganglion. They send out nerves to innervate the dorsal blood vessel, forming a cardio-aortic system, which controls the rate of heart beat, as well as having a secretory hormonal function.

Biological Clocks

Another function associated with the protocerebrum is time-keeping. Insect activity is timed with the daily light/dark cycle - the circadian cycle ('ciracdian' means 'about a day', the exact time being set each day according to environmental cues such as the length of daylight). This timing is due to internal clocks within the insect, which update themselves according to external cues from the environment (zeitgeibers or time-givers) such as the number of hours of light and dark. (This resetting by use of external signals enables the insect to adjust to different local conditions depending, for example, on latitude). Many body parts and organs have their own circadian clocks, indeed each cell appears capable of keeping time, but these appear to be set and synchronised by a central master clock, which resides in the protocerebrum and is both neural and hormonal. In some insects, a master clock is found in each optic lobe, which makes sense as these process light signals. There is also a daily movement of screening pigments in the ommatidia of the compound eye, as the insect adjusts to night-time darkness by increasing the sensitivity of its retina (it will continue to do this at the correct time for days when kept in constant light or dark for several days, so the response is coordinated, in part, by a central clock). Severing of the optic lobes prevents these clocks from synchronising bodily activities. In other species, however, the clock is only abolished if the brain is cut in two, which suggests that it may reside in the central body.

Deutocerebrum

This consists of two nerve centres - the main antennal lobe (AL) and the smaller antennal mechanosensory and motor centre (AMMC) or dorsal lobe. The AL receives inputs from the third (terminal) antennal segment (the flagellum, which is made-up of sub-segments called flagellomeres) via the antennal nerves. It contains from less than 10 to more than 200 sub-centres called glomeruli (singular glomerulus). Inputs to the AL appear to be mainly or exclusively from chemoreceptors (i.e. chemical sensors - olfactory and gustatory, smell and taste) on the flagellum. Each antenna sends signals to the AL on the same side of the head (ipsilateral pathways) although some may also send signals to the AL on the opposite side (contralateral pathways).

Each glomerulus is a region of neuropil (nerve cell processes and synapses) where computations occur. It is thought that each glomerulus may, in some species at least, receive inputs from a specific class of receptor (sensor) on the antenna. For example, in the males of some species there is a specially large glomerulus, called the macroglomerular complex (MGC) which receives inputs from pheromone olfactory sensors on the antenna.

The AL does not receive one input line from each chemoreceptor, as sensors of the same type converge - their axons fuse into a smaller number of axons in the antennal nerve (typically inputs from 15 sensors are combined, a 15:1 ratio). These sensory input axons, and also input axons from the CB of the protocerebrum, synapse with local interneurones within the AL (amacrine cells). Outputs from the AL are carried along the axons of output neurons to the MB of the protocerebrum.

The AMMC receives mechanosensory inputs from mechanosensors (mechanoreceptors)on the first two antennal segments (scape and pedicel) via the antennal nerves. It also sends motor outputs to the muscles of the scape. It also receives inputs from mechanosensors on the labial palps, some tegument (body wall) mechanosensors, and some inputs from the flagellum (possibly from the mechanosensors found on the flagellum). The antennal nerve is therefore a mixed nerve - containing both sensory and motor axons. Some of the antennal mechanoreceptors also send outputs to the SOG, the protocerebrum and the thoracic ganglia.

Tritocerebrum and Stomatogastric System

The frontal ganglion (FG) is an additional free and single (unpaired) median (median = in the midline) ganglion that is connected by a pair of bilateral connectives to the tritocerebrum. A single medial recurrent nerve runs back up to a ganglion situated beneath and behind the supraoesophageal ganglion. This ganglion may be called the stomachic ganglion or the hypocerebral ganglion (HG). In the locust, the HG sends out one pair of outer oesophageal nerves (and one pair of inner oesophageal nerves (ventricular nerves). Each of the latter terminates in a ventricular ganglion (ingluvial ganglion) on the crop of the foregut. These then control crop movements. In Dytiscus, it has been shown that the FG also controls swallowing. Thus, the tritocerebrum and frontal ganglion control the foregut, forming the stomatogastric system. The tritocerebrum also innervates the labrum.

Suboesophageal Ganglion

The suboesophageal ganglion (SOG) and the segmental ganglia of the double ventral nerve-cord each send out pairs of nerves, one of which innervates the pair of spiracles on that segment and so help regulate breathing. (In some insects the segmental ganglia are absent, e.g. in Dytiscus, in which case the lateral abdominal nerves send out nerves to innervate the spiracles). The SOG is a composite ganglion, formed by fusion of the ganglia from the mandibular, maxillary and labial segments of the head and the SOG also sends out nerves to the mouthparts (mandibles, palps, etc.) and so controls feeding behaviors.

The Ventral Nerve Cord

From the suboesophageal ganglion two connectives or nerve cords run back along the ventral side (underside) of the insect. These connect to the thoracic ganglion of the first thoracic segment, T1, which is actually a pair of ganglia, more-or-less fused into a single structure. T1 then gives off two connectives to the second thoracic ganglion, T2 and the sequence continues with a chain of connected ganglia running throughout the length of the insect, in the basic plan. Thus, we say that insects have a double ganglionated ventral nerve cord (VNC). Each ganglion functions as a local processor, regulating the functions of its body segment. The thoracic ganglia are especially well-developed as they have to carry out complex computations to generate patterns of movement in the legs and wings. These output patterns or central programs are contained in the ganglia, but the brain is normally required to switch them on and off. Sensory inputs have little effect on the basic patterns, but do modify them. For example, stress sensors in the wings feedback information to allow fine-adjustments to the wings and control of the angle of attack and wing-twisting. Typically, however, the basic pattern of movement is pre-coded.

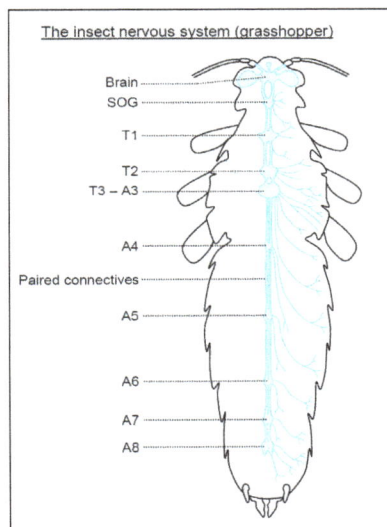

The exact arrangement of ganglia is grasshopper, left, the first three abdominal ganglia (A1, A2 and A3) are fused with the third thoracic ganglion (T3) to form a composite ganglion. The final abdominal ganglion is often composite. It is debatable how many abdominal segments there are, as the last few are modified and reduced, but generally there are 9-10, and the ganglia of these segments are fused with that of A8, again forming a composite ganglion ('A8'). Each ganglion gives off nerves to the various structures in its body segment. However, complications arise as ganglia may receive inputs from certain other segments too.

In flies ganglia are highly fused. Typically the three thoracic ganglia and the first four abdominal ganglia are fused together, into a single ganglion in the thorax. The remaining 5 (or so) abdominal ganglia are also fused into a single abdominal ganglion. The connectives of the nerve cord between the composite thoracic and composite abdominal ganglia then give off pairs of nerves to those abdominal segments lacking a regional ganglion. Such fusion of ganglia concentrates processing power where it is needed and reduces the time wasted by sending signals up and down the nerve cord between ganglia that may need to cooperate.

Learning, Memory and Intelligence in Insects

Although the mushroom bodies of the brain have been shown to be involved in learning, ganglia other than the brain are also capable of learning. Learning has been demonstrated in decapitated cockroaches. If a headless cockroach is wired so that one of its legs receives an electric shock when lowered, then it will learn to avoid the shocks by keeping the leg raised. (This is a classic experiment). The thoracic ganglia are responsible for controlling leg movement, and it is these ganglia that learn the new behavior. Intact cockroaches have a preference for darkness if given a choice between the illuminated half of a chamber and the darkened half. However, if they receive electric shocks in the darkened half, then they will learn to avoid the dark-half and remain in the light.

Habituation: An isolated cockroach leg exhibits another phenomenon related to learning - sensory habituation. If a touch-sensitive bristle is stimulated on an isolated leg and the activity in the leg nerve recorded, then it will be found that the strength of the stimulus diminishes if the stimulus is repeated rapidly, or sustained. This is due to fatigue in the periphery nerve and sensor, and this ensures that insects respond to changes in the environment, and learn to ignore persistent stimuli that are of no relevance. For example, body lice prefer rough fabrics, like wool, to smooth fabrics like silk. When crossing from wool to silk they will keep turning as they try to find their way back onto the wool. However, if they fail to find the wool (say if it is removed) then this behavior stops as the lice learn to make do - they habituate. Habituated insects are still responsive to changes in the stimulus, however. Thus, if the texture of the fabric for our louse on silk changes again, then we expect it will respond again, and habituate again if necessary.

Learning one's way about - route learning: Some insects, such as ants and cockroaches, are capable of learning the route of a maze, if the maze exit leads to reward, or if escaping from the maze avoids punishment. A cockroach will navigate a maze to it home-pot, so long as the home-pot contains recognisable cockroach odours. On subsequent trials it will reach its home-pot with increasing ease as it gradually learns the route. When compared to rats, ants learn mazes at half the speed: a rat will master a maze after about 15 runs, an ant after about 30, though the ants still make a few errors. (Not bad when you consider how tiny the ant brain is.). However, if a change is made to the maze, such as reversing the pattern, then rats learn the 'new' maze more rapidly than before, recognising the similarity in the patterns and transferring their previous learning to the new situation (transfer learning). In contrast, the ant starts all-over again, treating the maze as entirely new. Thus ants have little or no capacity for transfer learning, that is they seem unable to apply what they have learnt to a novel situation.

One of the most impressive feats of insect learning is locality learning: In addition to learning routes, insects can recognise the locality in which their nest is situated, or in which food is found. This involves exploratory learning- the insect will typically fly around a bit after leaving the nest, learning the position of many landmarks very rapidly and then leave. This is latent learning, meaning that a period of time elapses between learning and reward. The reward occurs when the insect returns home and locates its nest. This can be demonstrated by experiment. For example, in classic experiments on the beewolf, Philanthus triangulum a type of digger wasp, which brings back food to its developing young in the nest, the wasp will learn to recognise a ring of pine cones placed around the nest entrance and that the nest hole is the centre of this ring. If the pine cones are displaced a few centimetres when the wasp leaves, however, it will return to the centre of the ring of pine cones (only its nest is no longer there.).Generally, however, these insects are only temporarily confused by changes to one or two landmarks, falling back on other landmarks further from the nest (and perhaps other cues like smell?).

Associative learning: It occurs when a stimulus, irrelevant by itself, is made relevant by pairing it with something meaningful, like a food reward and the animal learns to identify the hitherto unconnected stimulus with food. For example, a cockroach can be rewarded by being given food when presented with a particular, but unrelated, odour. It will then learn to associate that odour with food and be attracted to it when foraging. Bees can be trained to associate a particular colour or pattern 9though their perception of shape is limited) with food.

Short-term and long-term memories: When an insect has just completed a task, learning is abolished if a new task is undertaken immediately after. Learning requires a latent period of rest in-between activities. To some extent the same is true of humans - learning becomes greatly enhanced if breaks are taken every 20 minutes or so during study. It is during these rest periods that the brain processes the information (often subconsciously in humans) and the appropriate neural pathways are reinforced. [Dreams in humans are especially curious - if activities are undertaken straight after waking then dreams are often very easily forgotten, but if a few minutes are spent reflecting on them they may be remembered more easily]. This can be explained by the existence of short-term memory (STM) and long-term memory (LTM). The nervous system requires time to process information in short-term memory, making sense of it and discarding information deemed irrelevant and transferring more useful information into long-term memory. In insects LTM typically retains information for several days.

Do Insects Sleep? Hard to say: Insects certainly exhibit circadian cycles. Insects enter a dormant or semi-dormant state when chilled. This may occur seasonally, during winter, or daily at nightfall. At nighttime insects may assume a particular posture in which to rest, bees may hold onto vegetation with their tightly clamped jaws, forexample. They may also return to the same resting spot each night. During these quiescent states the body temperature drops and energy is conserved. Interestingly, learning increases when insects are given rest between tasks or training sessions, as already mentioned, but also if they are chilled during these rest periods. Could it be that when chilled into torpor at nighttime insects are forming memories? It has been suggested that two chief functions of sleep are: energy conservation and memory formation. In this case the insect nighttime torpor is not so very different. Many insects are, of course, nocturnal. Cockroaches are nocturnal and learn better at night than during the day. It has been suggested that this encourages them to remember useful things like lessons learned during nighttime foraging.

Temporal learning: It occurs in insects. As we have seen insects can measure time, by the use of internal body-clocks, and they can also compensate for the changing position of the Sun in the sky when they use the Sun to navigate. To navigate successfully by the Sun they must know what time it is, in the sense of what portion of daylight has passed. Observations have shown that this may require learning, with young insects making mistakes by assuming that the Sun stays fixed in the sky.

Insight learning: It is a higher form of learning, similar to transfer learning, in that it takes prior learning and applies that to a new problem. One example is tool use. Some insects use tools, for example, some digger wasps fill in their nest burrow once the young is mature and ready to pupate (and so requires protection but no more food). The female may then hold a small 'pebble' in her mandibles and use it to pound-down the earth, and then discard the pebble. However, this behavior is not intelligent in the sense that the insect reasoned a solution, rather it is inherited genetically and is instinctive and so not true insight learning.

The Peripheral Nervous System

Peripheral nerves may be sensory or motory, but in insects are generally mixed. For example, the antennal nerve carries many sensory fibres conveying inputs from the many antennal sensors to the brain, but it also contains some motor fibres carrying output signals to the muscles in the base of the antenna. Animal nerve fibres, which are usually the axons of neurons, are typically wrapped by insulating glial cells. In nonmyelinated axons, the wrapping may be a simple sheath that loosely invests a group of axons. One function of this sheath is to ensure that the axons are bathed in a suitable salt solution necessary for them to conduct impulses. In myelinated axons, however, each individual axon is tightly wrapped in its own insulating sheath of material called myelin, which folds tightly around the axon several times, except at exposed regions (such as the nodes of Ranvier in mammals, although these may be simple pores in other animals). This more advanced type of wrapping is insulating and serves to speed-up nerve transmission. Students of vertebrate zoology often mistakenly believe that only vertebrates possess myelinated axons (in addition to unmyelinated axons which also occur in vertebrates). Myelinated axons occur in many invertebrates. Indeed, the fastest nerve transmission in the animal kingdom is seen in the myelinated axons of certain shrimps, which conduct signals at about 200 m/s. Insects have an intermediate form of insulation, which is like the myelin sheath, except that the myelin is wound loosely, leaving fluid-filled spaces between the layers. Such a nerve fibre is intermediate between unmyelinated and myelinated and is called a tunicated nerve fibre.

Respiratory System of Insects

An insect's respiratory system is the biological system with which it introduces respiratory gases to its interior and performs gas exchange.

Air enters the respiratory systems of insects through a series of external openings called spiracles. These external openings, which act as muscular valves in some insects, lead to the internal respiratory system, a densely networked array of tubes called tracheae. This network of transverse and longitudinal tracheae equalizes pressure throughout the system.

It is responsible for delivering sufficient oxygen (O_2) to all cells of the body and for removing carbon dioxide (CO_2) that is produced as a waste product of cellular respiration. The respiratory system of insects (and many other arthropods) is separate from the circulatory system.

Structure of the Spiracle

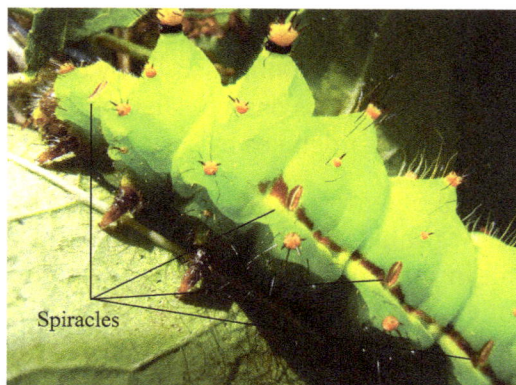

Indian moon moth (*Actias selene*) with some of the spiracles identified.

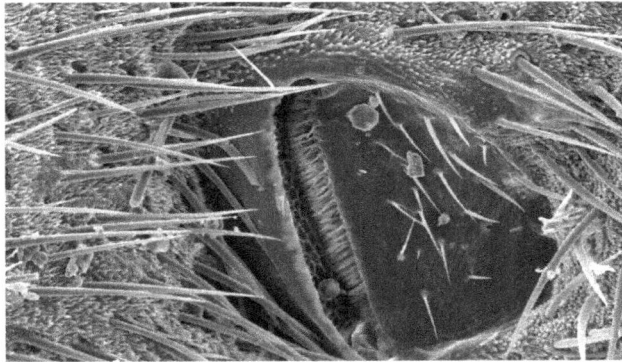

Scanning electron micrograph of a cricket spiracle valve.

Insects have spiracles on their exoskeletons to allow air to enter the trachea. In insects, the tracheal tubes primarily deliver oxygen directly into the insects' tissues. The spiracles can be opened and closed in an efficient manner to reduce water loss. This is done by contracting closer muscles surrounding the spiracle. In order to open, the muscle relaxes. The closer muscle is controlled by the central nervous system but can also react to localized chemical stimuli. Several aquatic insects have similar or alternative closing methods to prevent water from entering the trachea. Spiracles may also be surrounded by hairs to minimize bulk air movement around the opening, and thus minimize water loss.

The spiracles are located laterally along the thorax and abdomen of most insects—usually one pair of spiracles per body segment. Air flow is regulated by small muscles that operate one or two flap-like valves within each spiracle—contracting to close the spiracle, or relaxing to open it.

Structure of the Tracheae

After passing through a spiracle, air enters a longitudinal tracheal trunk, eventually diffusing throughout a complex, branching network of tracheal tubes that subdivides into smaller and smaller diameters and reaches every part of the body. At the end of each tracheal branch, a special cell (the tracheole) provides a thin, moist interface for the exchange of gasses between atmospheric air and a living cell. Oxygen in the tracheal tube first dissolves in the liquid of the tracheole and then diffuses across the cell membrane into the cytoplasm of an adjacent cell. At the same time, carbon dioxide, produced as a waste product of cellular respiration, diffuses out of the cell and, eventually, out of the body through the tracheal system.

Each tracheal tube develops as an invagination of the ectoderm during embryonic development. To prevent its collapse under pressure, a thin, reinforcing "wire" of cuticle (the taenidia) winds spirally through the membranous wall. This design (similar in structure to a heater hose on an automobile or an exhaust duct on a clothes dryer) gives tracheal tubes the ability to flex and stretch without developing kinks that might restrict air flow.

The absence of taenidia in certain parts of the tracheal system allows the formation of collapsible air sacs, balloon-like structures that may store a reserve of air. In dry terrestrial environments, this temporary air supply allows an insect to conserve water by closing its spiracles during periods of high evaporative stress. Aquatic insects consume the stored air while under water or use it to regulate buoyancy. During a molt, air sacs fill and enlarge as the insect breaks free of the old exoskeleton and expands a new one. Between molts, the air sacs provide room for new growth—shrinking in volume as they are compressed by expansion of internal organs.

Small insects rely almost exclusively on passive diffusion and physical activity for the movement of gasses within the tracheal system. However, larger insects may require active ventilation of the tracheal system (especially when active or under heat stress). They accomplish this by opening some spiracles and closing others while using abdominal muscles to alternately expand and contract body volume. Although these pulsating movements flush air from one end of the body to the other through the longitudinal tracheal trunks, diffusion is still important for distributing oxygen to individual cells through the network of smaller tracheal tubes. In fact, the rate of gas diffusion is regarded as one of the main limiting factors (along with weight of the exoskeleton) that prevents real insects from growing as large as the ones we see in horror movies. Periods in Earth's ancient history, however, such as the Carboniferous, featured much higher oxygen levels (up to 35%) that allowed horror movie sized insects, such as meganeura, along with arachnids, to exist.

Theoretical Models

Insects were once believed to exchange gases with the environment continuously by the simple diffusion of gases into the tracheal system. More recently, large variation in insect ventilatory patterns have been documented, suggesting that insect respiration is highly variable. Some small insects do demonstrate continuous respiration and may lack muscular control of the spiracles. Others, however, utilize muscular contraction of the abdomen along with coordinated spiracle contraction and relaxation to generate cyclical gas exchange patterns and to reduce water loss into the atmosphere. The most extreme form of these patterns is termed discontinuous gas exchange cycles (DGC). Recent modeling has described the mechanism of air transport in cyclic gas exchange computationally and analytically.

Insects Circulatory System

Insects have an open circulatory system. This means that the internal organs and tissues are bathed in hemolymph, which is propelled actively to all internal surfaces by specialized pumps, pressure pulses, and body movements and is directed by vessels, tubes, and diaphragms. Without such constant bathing, tissues would die. The internal organs and tissues depend on the circulatory system for the delivery of nutrients, and to carry away excretion products, and as the pathway by which hormone messengers coordinate development and other processes.

Gas exchange in insects occurs via the tracheal system, which supplies all internal organs with tracheole tubules from spiracular openings in the body wall of terrestrial insects or from gill structures in aquatic insects. However, the hemolymph has the capacity to dissolve carbon dioxide gas in the form of bicarbonate ions. A few insects live in low oxygen environments and have a type of hemoglobin that binds oxygen at very low partial pressures, but for the most part oxygen is supplied and carbon dioxide is removed by ventilation through the tracheal system. Besides the functions already mentioned, the circulatory system provides a medium in which battles are fought between the insect host and a myriad of invading disease microorganisms, including viruses, bacteria, fungi, and insect parasites. Principal participants in these interactions are the blood cells or hemocytes.

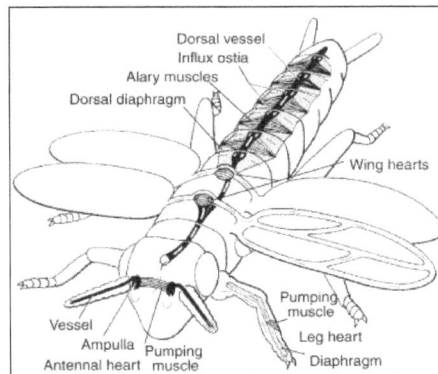

Delivery of the hemolymph to all tissues is so vital that a number of structures have evolved to ensure complete circulation. Principal circulatory organ is the dorsal vessel which is supported by the underlying dorsal diaphragm and the ventral diaphragm. Circulation in the appendages is effectuated by accessory pulsatile organs.

While maintaining the body tissues, the circulatory system is the medium in which homeostasis is ensured, including the regulation of pH and inorganic ions, as well as the maintenance of proper levels of amino acids, proteins, nucleic acids, carbohydrates, and lipids. Any change in the hemolymph quickly affects all organs bathed. The time for complete mixing of the hemolymph depends on the size of the insect, but it can be up to 5 min in a resting adult cockroach weighing about a gram. Any substance injected into a healthy insect will eventually appear at the extreme ends of all appendages in a few minutes, emphasizing the efficiency of the delivery mechanisms, which can be marvels of microhydraulic engineering.

Dorsal Vessel

The principal organ of hemolymph propulsion is the dorsal vessel, or at least it is the most visible organ specialized in hemolymph movement in insects. It forms a hollow tube which runs along the midline for the whole length of the body. Contraction of its circular musculature results in a contractile stroke (called systole), whereas elastic connective tissue strands, which connect the dorsal vessel to the body wall, are responsible for dilation and opening of the vessel (called diastole).

The dorsal vessel of most insects is not uniform in its course through the body and shows a differentiation into two regions which is reflected by their traditional denomination: the posterior part in the abdomen is referred to as the "heart," whereas the anterior part in the thorax and the head is the "aorta." Both terms are borrowed from better-known vertebrate structures and give an inaccurate impression of the different roles of those structures in insects. In phylogenetically ancestral insects, the flow of hemolymph in the dorsal vessel is bidirectional: upon contraction hemolymph flows toward the head in the anterior part and simultaneously toward the rear end in the posterior part of the dorsal vessel. The posterior directed flow is caused by an intracardiac valve in the abdomen and supplies the long caudal appendages (cerci and terminal filum), for example in some apterygotes and mayflies. However, in most insects the hemolymph flow is unidirectional in that the contraction of the dorsal vessel begins at the posterior end and advances forward as a peristaltic wave. In the pupae and adults of Coleoptera, Diptera, and Lepidoptera there is a regular change in the direction of the contraction wave, termed heartbeat reversal. Contraction waves of the dorsal vessel toward the head alternate with waves toward the rear of the body; between these contractions are short phases of rest.

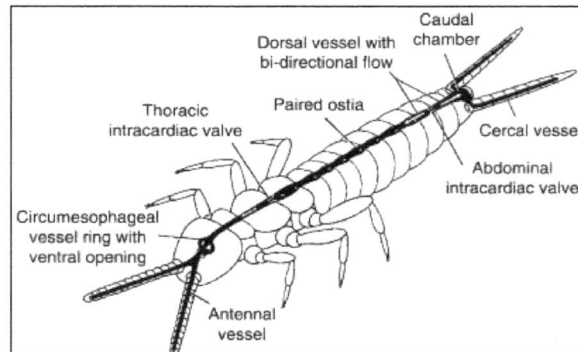

In the primitive insect Campodea (Diplura) the dorsal vessel exhibits a bidirectional flow. This enables the supply of the antennae and the cercal appendages by vessels connected to the dorsal vessel.

Hemolymph flows into the dorsal vessel through a pair of lateral openings (called ostia) in each abdominal segment. Valves which project into the vessel lumen close the ostia at the systolic contraction. Hemolymph emerges from the dorsal vessel at the open anterior end, and in insects with heartbeat reversal it usually also flows out the open posterior end. In Lepidoptera, there are two-way ostia which permit influx during forward phases and outflow during backward phases. The anterior end of the dorsal vessel opens just beneath or in front of the brain. This arrangement ensures a constant supply of nutrients and removal of waste products to and from the brain mass. In addition, the dorsal vessel is often intimately associated with the retrocerebral nervous system (including the hypocerebral ganglion, corpora cardiaca, and corpora allata complex) just behind the brain, which delivers neu-rohormones and possibly other hormones into the aorta at specialized release sites. Some insects have in addition outflow openings which are paired or unpaired and located more ventrally. In cockroaches, man-tids, and some orthopterans, the outflow openings are outfitted with sphincter-like valves and associated with segmental vessels laterally diverging from the heart. These vessels are formations of connective tissue with no inherent musculature, thus providing a simple channel to ensure lateral perfusion of the pericardial sinus.

The dorsal vessel is comprised of two rows of opposing pairs of muscle cells (collectively called myocardium). The cells of each row are offset, so that they cause a spiral-like peristaltic wave of contraction. The myocardium in all insects is spontaneously active, usually beginning in the embryonic stages. This type of heart control is termed myogenic, because the electrical activity underlying contractions arises in the myocardium itself. This is in contrast to a neuro-genic heart control present in crustaceans, such as crabs and lobsters, in which a barrage of nervous impulses drives the heartbeat from a discrete cardiac ganglion center. In insects, the myogenic pacemaker may be neurally or hormonally modulated. The basal heartbeat rate of most insects is around 60 beats min-1 at room temperature and at rest (e.g., in the American cockroach, Periplaneta americana, and the locust, Locusta migratoria). In adult house flies (Musca domes-tica), however, it ranges between 300 or more beats per minute during flight to zero beats for a while when at rest. The central nervous system of the adult house fly is composed of the brain and a thoracic ganglion mass, but lacks abdominal ganglia. Because of this unusual anatomy, the dorsal vessel of the abdomen can be separated from central nervous system input by experimentally severing the thorax from the abdomen. After this operation, the heartbeat of the fly becomes quite regular at around

60 beats min- 1. This indicates that the heart of the house fly is innervated by both inhibitory and excitatory motor neurons from the central nervous system.

Dorsal Diaphragm

The dorsal diaphragm is a fenestrated membrane which separates the upper pericardial sinus from the lower perivisceral sinus. It consists of connective tissue with associated muscles which are called alary muscles because in some insects, for example cockroaches, they resemble wings projecting laterally from the dorsal vessel of each abdominal segment. Although sometimes mistakenly thought to play a key role in heartbeat, the alary muscles are more properly called muscles of the dorsal diaphragm. Whereas the myocardium is specialized to contract rapidly and constantly, the ultrastructure of the alary muscles indicates infrequent and slow contractions. The muscles associated with the dorsal diaphragm may likewise be arranged in a loose network, as in Lepidoptera, or arranged like a weave surrounding the dorsal vessel, as in some Diptera; the functional role of these muscles is difficult to determine.

Ventral Diaphragm

The ventral diaphragm plays a prominent role in perfusing the ventral nerve cord of insects. Nearly 40 years ago Glenn Richards surveyed the ventral diaphragms in insects and found that insects with a well-defined ventral nerve cord in the abdomen also have a well-developed ventral diaphragm. In contrast, insects with the ventral nerve cord condensed into a complex ganglion structure in the thorax invariably lack a ventral diaphragm. This correlation suggests that the role of the ventral diaphragm is inexorably tied to perfusion of the ventral nerve cord in the abdomen. When present, the ventral diaphragm loosely defines a perineural sinus below and the perivis-ceral sinus above containing the gut. In some insects, the ventral diaphragm is a strong muscular structure with a great deal of contractile activity. The activity of the ventral diaphragm is dictated by innerva-tion from the central nervous system. In some large flying insects, the ventral diaphragm assists in hemolymph flow during thermoreg-ulation by facilitating the removal of warm hemolymph from the hot thoracic muscles to the abdomen for cooling. The intimate association between the ventral diaphragm in insects and perfusion of the ventral nerve cord is strengthened by considering the structure in cockroaches that takes the place of a proper diaphragm. In these insects, four stripes of muscle, together called hyperneural muscle,are near the back of each of the abdominal ganglia, and contract slowly but not in a rhythmic order. The muscles are electrically inex-citable, which means that they do not contract myogenically, as the myocardium does, but instead are neurally driven by motor neurons located in the ventral ganglia. Thus each of the ventral nerve cords in cockroaches has its own muscle supply that pulls it back and forth along the midline of the abdomen upon demand thereby increasing the contact and mixing between the ganglia and the hemolymph.

Accessory Pulsatile Organs

The dorsal vessel and the two large diaphragms are responsible for pumping hemolymph through the main body cavity. However, these organs are incapable of achieving circulation in body appendages such as the antennae, legs, wings, and various long abdominal processes (e.g., cerci and ovipositors). To supply these appendages, insects rely on special, small circulatory pumps, known collectively as accessory pulsatile organs or accessory hearts. As a rule, accessory hearts are separate from the dorsal vessel and function autonomously. The pumping organ is generally located at the base of the appendage and is connected to vessels or special diaphragms which guide the flow of hemolymph through the appendage. Accessory pulsatile organs are present in an astounding array of functional constructions. They are evolutionary novelties of higher insects and are absent in the phylogenetically ancestral insects. Antennal circulatory organs are nearly universal in insects, lacking only in groups with extremely short antennae, such as fleas and lice. The ancestral state of antennal circulation in insects is antennal vessels connected to the dorsal vessel as in certain apterygotes. The connection between the two vessels was probably lost during the further evolution of insects and replaced by autonomous pulsatile organs at the base of the antennal vessels. The anatomy of these organs differs widely in various insects. The best investigated antennal heart in terms of morphology and physiology is that of the American cockroach. Remarkably, the antennal heart of the cockroach functions not only as a circulatory pump but also as a neurohemal organ: hormones released into the ampulla lumen are pumped into the antennae where they most likely modulate the sensitivity of the numerous sensilla.

In leg circulatory organs, the flow of hemolymph is guided by longitudinal diaphragms instead of vessels. A diaphragm of connective tissue divides the inner cavity of each leg into two channels permitting a counter-current flow. Some insects have pulsatile leg organs with muscular attachment to the diaphragm. When the muscle contracts, one channel is compressed forcing hemolymph toward the thorax; at the same time, the other channel expands drawing hemolymph into the leg. Other insects utilize changes in the volume of an elastic tracheal sac in the legs as the driving force for hemolymph exchange. A common misconception is that insect wings are dead cuticular structures. However, the veins of the wings are filled with hemolymph to maintain living tissues, such as nerves and tracheae. Circulation is achieved by pumping organs in the thorax, which in ancestral winged insects are ampullary enlargements of the dorsal vessel. The wing hearts of most holometabolan insects, however, are muscular diaphragms which are separate from the dorsal vessel and which are either paired or unpaired. Recently, a further vital function of the wing hearts was discovered in Drosophila, namely that they are essential for the proper maturation of wings. Toward the completion of the wing formation process, they suck epidermal cells out of the wings which are necessary before the two cuticular surfaces of each wing bond together. Flies lacking functional wing hearts never develop flight ability. Finally it should be noted that accessory hearts may also partake in the hydraulic movements of body appendages. For example, the long sucking mouthparts of butterflies are uncoiled by action of special pumping organs in the head, and the lamellae of the antenna in scarabaeid beetles are spread out by action of the antennal heart.

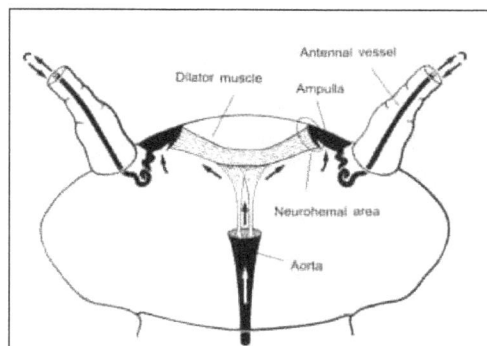

The antennal heart of the cockroach, Periplaneta americana, consists of a pulsatile ampulla at the base of each anten-nal vessel. The two ampullae expand by action of the interconnecting dilator muscle. Hemolymph rushes into each ampulla through a small ostium. When the muscle relaxes, the ampullae collapse by their own elasticity forcing the hemolymph into the antennal vessels. The hemolymph is conveyed to the end of each vessel and empties into the lumen of the antenna before returning to the head capsule; the two small muscles which extend to the anterior opening of the aorta are responsible for the suspension of the apparatus.

Extracardiac Pulsations

First described in 1971, extracardiac pulsations of insects are the simultaneous contractions of interseg-mental muscles, usually of the abdomen, that cause a sharp increase in the pressure in the insect body. The amount of movement accompanying each pulse is too small to be seen, but it can be readily mea-sured as a slight shortening or telescoping of the abdomen as measured from its tip. The extracardiac pulses should not be confused with larger overt movements of the abdomen, especially in bees and bumble bees, that accompany ventilation during times or high activity or exertion such as flight.

Either the extracardiac pulsations occur in coordination with openings of certain of the spiracles, and therefore can play a role in ventilation, or they occur when all the spiracles are tightly closed, hence affecting hemolymph movement. The extracardiac pulsations become suspended only in quiescent stages of insect development, such as during diapause, but they can be evoked immedi-ately upon disturbance or stimulation.

The extracardiac pulsations are driven by a part of the nervous system for which Karel Slama coined the name "coelopulse nervous system." The pressures induced by extracardiac pulsations are 100-500 times greater than pressures caused by contractions of the dorsal vessel and are trans-mitted by the hemolymph throughout the entire body of the insect, influencing hemolymph move-ment at some distance from the dorsal vessel and APO structures.

Tidal Flow of Hemolymph

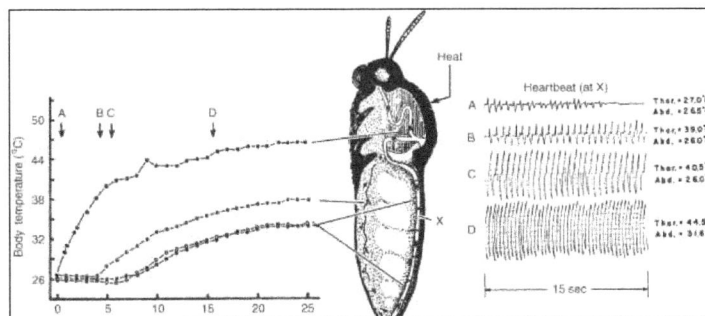

Control of thoracic temperature by central nervous control of dorsal vessel contractions during external heating of the thorax (heat). At the optimum temperature, hemolymph is pumped at maximum frequency and amplitude through the dorsal vessel to conduct heat from the thorax to the abdomen, where it is dissipated.

A special condition of the circulatory system exists in some large, high-performance flyers, such as some Lepidoptera, Diptera, and Hymenoptera. To keep body weight at a minimum the amount of hemolymph is reduced and the volume replaced by large tracheal sacs. The body is often divided into two hemocoel compartments by an anatomical constriction between thorax and abdomen and may be additionally separated by a valve. The hemolymph in these insects is not circulated in the classical sense but is transported back and forth between thorax and abdomen by heartbeat reversal. The shift in hemolymph flow causes an alternating periodic increase and decrease in hemolymph volume in both compartments with a compensatory volume change in the tracheal system. This leads to ventilation especially of the large elastic tracheal sacs. Furthermore, the hemolymph in the wings oscillates in all veins simultaneously in correlation with heartbeat reversal and the resulting volume changes in the wing tracheae. Lutz Wasserthal called this periodic exchange of air and hemolymph in these insects "tidal flow" of hemolymph.

Thermoregulation

The use of the hemolymph in thermoregulation of flying insects was firstly described by Bernd Heinrich. The optimum temperature for flight muscle contraction in many insects, such the tobacco hornworm, Manduca sexta, is surprisingly high, up to 45 °C. Before this moth can fly, it must warm the thorax to near this temperature, which it accomplishes by means of a series of simultaneous isometric contractions of the antagonistic pairs of flight muscles that appear to the casual observer as "shivering," or vibrations of the wings.

A "thermometer" in the thoracic ganglia detects the proper temperature. When the thoracic temperature is below optimum, the central nervous system signals the dorsal vessel to circulate hemol-ymph slowly. When the thoracic temperature rises above optimum, the central nervous system brings about maximal amplitude and rate of heartbeat to drive hemolymph through the thoracic muscles. The increased hemolymph flow pulls heat away from the flight muscles in the thorax and eventually delivers hot hemolymph to the abdomen, where the heat is dissipated. Then relatively cool hemolymph is redelivered to the thoracic muscles by the dorsal vessel, completing the thermoregulation cycle. The warm hemolymph is then delivered to the head and percolates back past the ventral ganglia in the thorax to the abdomen, where the heat is dissipated. The cooler hemol-ymph is then delivered again to the thorax. The dorsal vessel and the very strong ventral diaphragm in the tobacco hornworm act together to move hemolymph. When the thorax is too warm, both the amplitude and the frequency of heartbeat contractions are increased, and the rate of delivery of hemolymph increases. When the thorax is too cool, amplitude and frequency of contraction of the dorsal vessel are decreased. The activity of the ventral diaphragm acts in concert with that of the dorsal vessel.

Thermoregulation of the flight muscles of the tobacco hornworm implies a sophisticated nervous control. The overall nervous control can be easily demonstrated by severing the ventral nerve cord between the thorax and abdomen. Then the moth can no longer thermoregu-late because the feedback loop of temperature detection by the thoracic ganglia has been destroyed, and control over ventral diaphragm and dorsal vessel contractions has been lost.

Autonomic Nervous System

The tidal flow of hemolymph, the extracardiac pulsations, heartbeat reversal, and thermoregulation all imply a very sophisticated control of circulation by the central nervous system. The central nervous system also plays a role in regulation of the respiratory system. The activities of circulatory and respiratory systems are coordinated by the central nervous system, perhaps to an extent not fully appreciated. It would be convenient and satisfying to be able to point out a particular part of the central nervous system and related peripheral nerves in insects that might comprise this regulatory system; however, beyond evidence that the meso- and/or metathoracic ganglia play a major role in some of these functions, entomologists know of no such discrete structure or structures, possibly because these interregulatory functions have been undertaken by different parts of the nervous system in different insects. It is known that insects have a number of regulatory mechanisms that can be recruited to achieve such control, from motor and sensory neurons to neurosecretory neurons and neurohormonal organs located throughout the insect hemocoel.

Insect Reproductive System

The reproductive organs of insects are similar in structure and function to those of vertebrates: a male's testes produce sperm and a female's ovaries produce eggs (ova). Both types of gametes are haploid and unicellular, but eggs are usually much larger in volume than sperm.

Most (but not all) insect species are bisexual and biparental — meaning that one egg from a female and one sperm from a male fuse (syngamy) to produce a diploid zygote. There are, however, some species that are able to reproduce by parthenogenesis, a form of asexual reproduction in which new individuals develop from an unfertilized egg (virgin birth). Some of these species alternate between sexual and asexual reproduction (not all generations produce males), while others are exclusively parthenogenetic (no males ever occur).

Sexual reproduction might well be the most important "adaptation" ever acquired by living organisms. It provides a mechanism for shuffling and recombining genetic information from two parents to create new ("hybrid") genotypes that can be tested in the fire of natural selection. Only phenotypes that withstand the "heat" can participate in the next round of reproduction.

External vs. Internal Fertilization

Stalked spermatophore of a collembolan.

As long as primitive arthropods lived in the water, their sperm could simply swim from the male's body to the female's body where fertilization could occur. But in order to adopt a terrestrial lifestyle, animals that engaged in such external fertilization had to protect their sperm from desiccation. The solution, still used today by myriapods and insects, was to encapsulate large numbers of sperm within a water-tight lipoprotein shell secreted by the male's accessory glands. These "packages" of sperm are known as spermatophores. In myriapods and primitive hexapods (e.g. Collembola), males leave spermatophores on the ground where they may be found and picked up by a passing female. Silverfish and bristletails have more elaborate courtship activities in which the male leads his mate to a freshly deposited spermatophore.

Today, all of the more "advanced" insects exhibit internal fertilization — males deposit their sperm inside a female's body during an act of copulation. This novel adaptation, which appeared soon after insects diverged from their myriapod-like ancestors, presumably ensured that more sperm found their way to a receptive female. But the genetic programming for spermatophore production still persists in most modern insects. After a male deposits his spermatophore inside a female's reproductive system, she digests the lipo-protein coat and uses it as a source of additional nutrition for her eggs. In some cases, the quality (or quantity) of this nuptial gift may even determine whether a female accepts or rejects the male's gametes.

Sex Determination

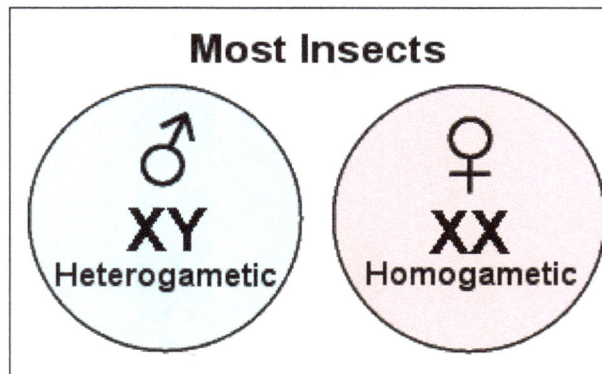

Like humans, most insects have a single pair of chromosomes that carry the genetic information for determining an individual's gender. If an embryo inherits a pair of "X" chromosomes, it will develop as a female; if it inherits one "X" and one "Y", it will develop as a male. The "XX" female is said to be homogametic; the "XY" male is heterogametic. In this case (as in humans) the male's contribution determines the offspring's gender. Some insect species have no "Y" chromosome at all — males have just one "X", and females have two. A similar condition is found in some parthenogenetic species of aphids in which "maleness" occurs through the loss (degeneration) of one chromosome during embryogenesis. In both cases, the males end up with an odd number of chromosomes ($2n-1$).

Trichoptera & Lepidoptera

♂
WW
Homogametic

♀
WZ
Heterogametic

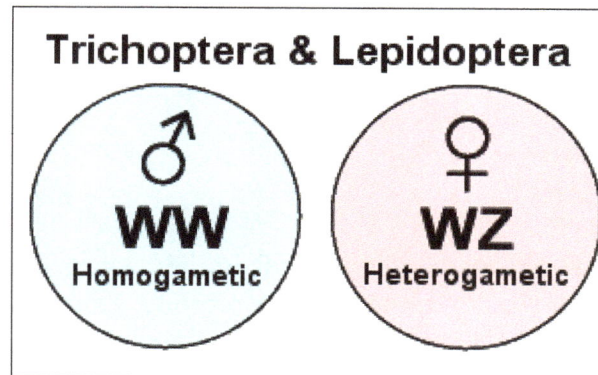

In Lepidoptera and Trichoptera, however, the homo and heterogametic sexes are reversed: females are heterogametic and males are homogametic. To distinguish this system from standard X-Y sex determination, these sex chromosomes are designated "W" and "Z" (instead of "X" and "Y"). Thus, a female butterfly is "WZ" and a male butterfly is "WW". In this case, the female's contribution determines the offspring's gender. Oddly, there is only one other group of organisms in the animal kingdom that has this pattern of sex determination.

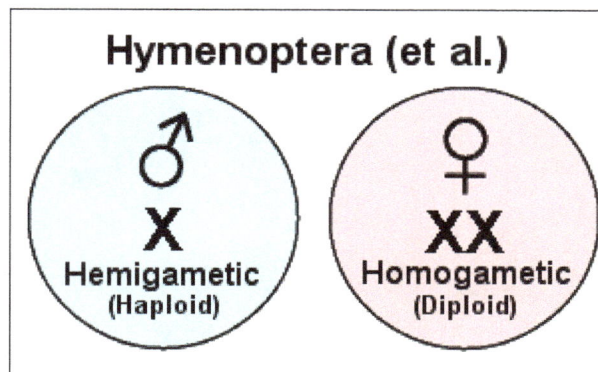

Hymenoptera (et al.)

♂
X
Hemigametic
(Haploid)

♀
XX
Homogametic
(Diploid)

A third method of sex determination, called haplo-diploidy, is found in all Hymenoptera, many Thysanoptera, some scale insects (Hemiptera/Homoptera), and a few weevils (Coleoptera). These insects have diploid, homogametic females ("XX"), but all of the males are haploid — they develop by parthenogenesis (asexually) from unfertilized eggs. Primary oocytes undergo meiosis to form haploid eggs, but meiosis is unnecessary in primary spermatocytes because the cells are already haploid. Unmated females can lay eggs that will develop into males. Once a female mates and receives sperm from a male, she has two options:

1. She can produce a female offspring by opening the valve at the base of her spermatheca to release sperm onto the egg as it passes through her oviduct, or

2. She can produce a male offspring by closing the spermathecal valve and preventing any sperm from reaching the egg.

Control over the gender of offspring has proven to be a useful adaptation for some insects. A biased sex ratio that favors females over males can reduce competition for limited food resources and increase the reproductive potential of the population. Bees, wasps, and ants form large colonies of queens and workers (all female) in which males are produced only sporadically as needed for reproduction.

Worm Physiology

Even though they don't have a skeleton and can't walk upright, worms do many of the same things people do to survive. They move around, eat, breathe, reproduce and defend themselves. They are sensitive to temperature, moisture, light and vibrations.

Annelids

Worms are annelids, derived from the word anulus meaning "ring." Worms are made up of joined, ringed segments. An adult redworm has between 200 to 400 circular rings. Try thinking of the giant sequoia trees and their rings. The tree rings grow around each other while the worm rings are stacked end on end.

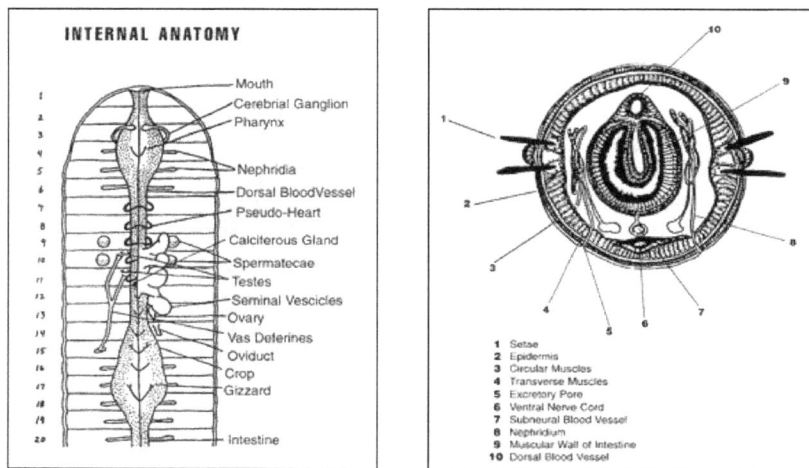

INTERNAL ANATOMY

Mouth
Cerebral Ganglion
Pharynx
Nephridia
Dorsal BloodVessel
Pseudo-Heart
Calciferous Gland
Spermatecae
Testes
Seminal Vescicles
Ovary
Vas Deferines
Oviduct
Crop
Gizzard
Intestine

1 Setae
2 Epidermis
3 Circular Muscles
4 Transverse Muscles
5 Excretory Pore
6 Ventral Nerve Cord
7 Subneural Blood Vessel
8 Nephridium
9 Muscular Wall of Intestine
10 Dorsal Blood Vessel

The cuticle is the worm's outermost body wall. Beneath the cuticle are:

1. The epidermis, which is like our skin,

2. A layer of nerve tissue which performs like our sense of touch,

3. Circular and longitudinal muscles for locomotion.

The epidermis contains many sensory cells that transmit information to the nerve tissue. Within the layer of nerve tissue are cells that forward sensory information to the worm's nerve cord and on to the cerebral ganglion, the worm's version of a brain.

Circular muscles create the worm's body rings. These muscles contract and expand, shortening and lengthening the worm's body. The longitudinal muscles run the length of the worm. Acting in concert these sets of muscles enable the worm to propel forward, backward and sideways. Moisture in their environment lubricates this locomotion.

Additionally, stiff hair-like protusions called setae stick out of almost every ring along each side of the worm. These protusions grip the soil, bedding or any material it is moving through so the worm doesn't just slip and slide randomly. The setae are made of chitin, the same substance that makes up our fingernails and the exoskeleton of many insects. The setae are very strong, assisting the worm as it moves through its environment. The setae help the worm to defend itself by gripping the soil when attacked by a hungry bird or other predator.

Brain and Nervous System

The cerebral ganglion, located at the front of the worm, serves as the brain. This nerve bundle is responsible for receiving external information such as light, heat, moisture and vibrations. The worm relies on the ganglion and a ventral nerve cord for sensory input from the world around them.

While we don't fully understand all the functions of the nervous system, it is believed that body functions such as reproduction and life cycles are regulated within the nerve ganglion.

Circulation

The circulatory system is powered by five pseudo-hearts. These hearts are merely valved chambers that regulate blood flow and produce a pulse. Branching off these hearts are both a dorsal (forward flow) and a ventral (backward flow) blood vessel. The dorsal and ventral vessels transport the blood, rich with oxygen and nutrients through the body. The circulatory system also transports urinary waste, which is diffused through the cutical, the outer covering, in each ringed segment. In other words, worms breathe and excrete urine through their skin.

Digestion/Gastrointestinal System

Running through the worm's body is the alimentary canal or gut. It starts at the mouth, called the buccal cavity, and moves to the back with the pharynx, esophagus, crop, gizzard, intestine and anus, respectively.

The buccal cavity contains specialized sensory cells which allow worms to locate food and minerals. The cells detect and recognize sucrose, glucose, quinine and saline chemicals from the environment. This allows them to identify and select the foods they eat.

The pharnyx works like a suction pump, drawing particles farther in from the mouth. The esophagus, which opens from the pharnyx as a narrow tube, leads to the crop. Worms and birds both use their crop as a food storage chamber.

Next is the gizzard, the food grinding chamber. It contains sandy grit from the soil to pulverize food into small particles, including leaf litter, mulch and soil organics.

The intestine is a tube going straight back to the anus, taking up almost two-thirds of the length of the worm. The intestine performs the final digestion and absorption of the life sustaining nutrients from the worm's food.

Enzyme Benefits

Many tiny organisms, bactria, fungi, actinomycetes, enzymes and protozoa live in the worm's gastrointestinal systems aiding digestion. They are microscopic and thrive by the hundreds of thousands within a single worm. These organisms assist in preparing the nutrients to be absorbed and utilized by the worm.

The worm produces numerous enzymes which aid in its own survival, including an insecticide and an antibiotic. These enzymes emulsify with mucus produced in the worm's gut and sheath the castings when expelled through the anus. Plants are able to absorb the insecticidal and antibiotic enzymes through their roots to further utlize them in the plants' ongoing battle to ward off insects and disease. The antibiotic enzymes also protect humans from harmful bacteria while working in the worm bin.

Respiration

Worms have no specialized respiratory organs. They breathe through their moist skin. Oxygen and carbon dioxide are diffused through the skin to and from the circulating blood stream. Lack of moisture in the worm's environment restricts the breathing process. Prolonged dryness will cause death by suffocation. Exposure to direct sunlight can lead to death in less than three minutes.

References

- Mohs, K.; McGee, I. (2007). Animal planet: the most extreme bugs (1st ed.). John Wiley & Sons. p. 35. ISBN 978-0-7879-8663-6

- Invertebrate, entry: newworldencyclopedia.org, Retrieved 23 June, 2019

- Völkel, R.; Eisner, M.; Weible, K.J. (2003). "Miniaturized imaging systems". Microelectronic Engineering. 67-68: 461–472. doi:10.1016/S0167-9317(03)00102-3. ISSN 0167-9317

- Insect-nervous-systems, BioTech: cronodon.com, Retrieved 23 July, 2019

- Circulatory-system-insects, insects: what-when-how.com, Retrieved 20 July, 2019

- Reproductive-system, bug-bytes: ncsu.edu, Retrieved 31 July, 2019

- Eggleton, P (2001). "Termites and trees: a review of recent advances in termite phylogenetics". Insectes Sociaux. 48: 187–193. doi:10.1007/pl00001766

- Vc-anatomy, composting: eulesstx.gov, Retrieved 1 May, 2019

4
Animal Cell Physiology

The biological branch which focuses on the activities which take place within cells is termed as cell physiology. Some of the processes studied within this field are cell adhesion, cell division and cell signaling. The topics elaborated in this chapter will help in gaining a better perspective about these areas of study within cell physiology.

Animal Cells

As with all of Earth's organisms, animals are built from microscopic structures called cells. Cells are the basic unit of life and these microscopic structures work together and perform all the necessary functions to keep an animal alive. There is an enormous range of animal cells. Each is adapted to a perform specific functions, such as carrying oxygen, contracting muscles, secreting mucus, or protecting organs.

The cells of animals are advanced and complex. Along with plants and fungi, the cells of animals are eukaryotic. Eukaryotic cells are relatively large cells with a nucleus and specialized structures called organelles.

Although animal cells can vary considerably depending on their purpose, there are some general characteristics that are common to all cells. These include structures such as the plasma membrane, cytoplasm, nucleus, mitochondria, and ribosomes.

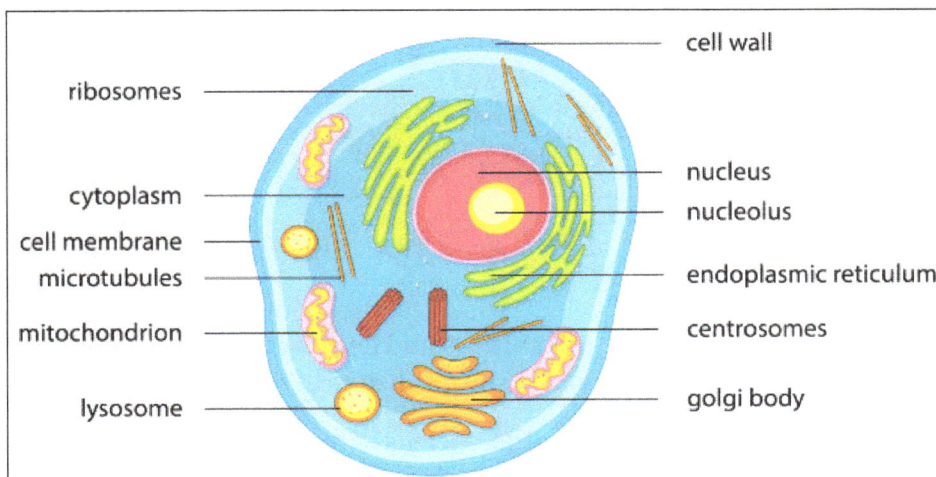

General Structure of an Animal Cell

Animal cells have a number of organelles and structures that perform specific functions for the cell. The huge variety of cells that have evolved to fulfill different purposes do not always have all the same organelles or structures, but in general terms, these are some of the structures you can expect to find in animal cells:

Plasma Membrane

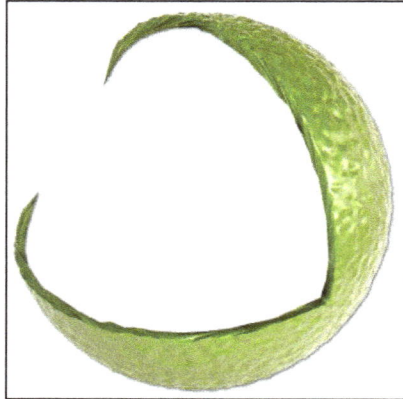

The plasma membrane is a porous membrane that surrounds an animal cell. It is responsible for regulating what moves in and out of a cell. The plasma membrane is made from a double layer of lipids. Extra compounds such as proteins and carbohydrates are embedded into the lipid membrane and perform roles such as receiving cellular signals and creating channels through the membrane.

Nucleus

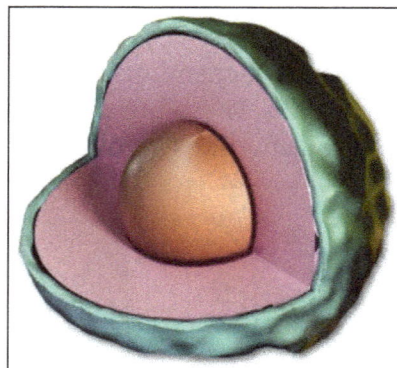

The cells of animals and plants almost always have a 'true' nucleus. A nucleus consists of a nuclear envelope, chromatin, and a nucleolus.

The nuclear envelope is made from two membranes and encapsulates the contents of the nucleus. The double membrane has numerous pores to allow substances to move in and out of the nucleus.

Inside the nuclear envelope, the majority of the nucleus is filled with chromatin. Chromatin contains the majority of a cell's DNA and condenses down to chromosomes as a cell divides. The nucleolus is the center core of the nucleus and produces organelles called ribosomes.

Cytoplasm

The cytoplasm is the internal area of an animal cell that isn't occupied by an organelle or nucleus. It consists of a jelly-like substance called 'cytosol' and allows organelles and cellular substances to move around the cell as needed.

Endoplasmic Reticulum (Er)

The endoplasmic reticulum is a network of membranes found within almost all eukaryotic cells. The membranes are connected to the membrane of the cell's nucleus and are important for many cellular processes such as protein production and the metabolism of lipids and carbohydrates.

The endoplasmic reticulum includes both the smooth ER and the rough ER. The smooth ER is a smooth membrane and has no ribosomes, whereas the rough ER has ribosomes that are used to produce proteins.

Mitochondria

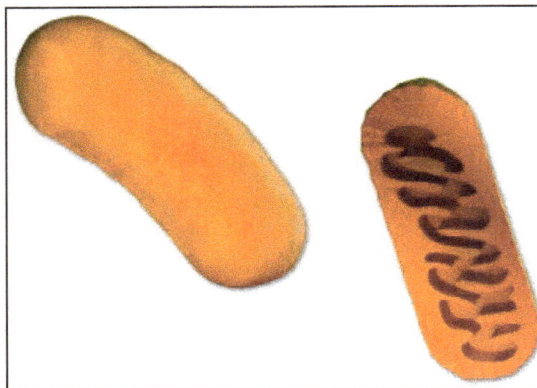

Mitochondria are one of the most important of all organelles. They are the site of cellular respiration – the process that breaks down sugars and other compounds into cellular energy. It is in the mitochondria where oxygen is used and CO_2 is produced as a byproduct of respiration.

Golgi Apparatus

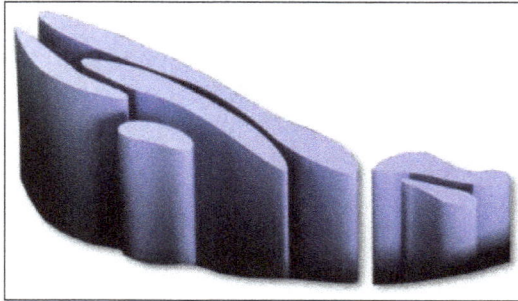

The golgi apparatus (or golgi body) is another set of membranes found within the cell but is not attached to the nucleus of the cell. It serves many important functions including modifying proteins and lipids and transporting cellular substances out of the cell.

Ribosomes

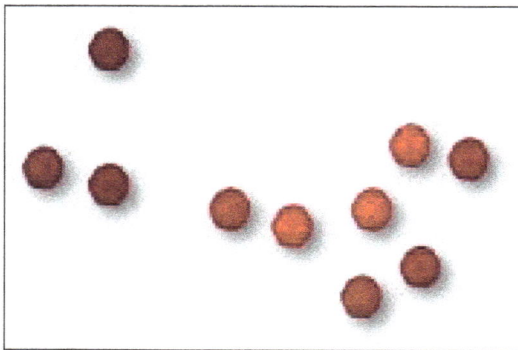

Ribosomes are involved in the process of creating proteins. They can be either attached to the endoplasmic reticulum or floating freely in the cell's cytoplasm.

Peroxisomes

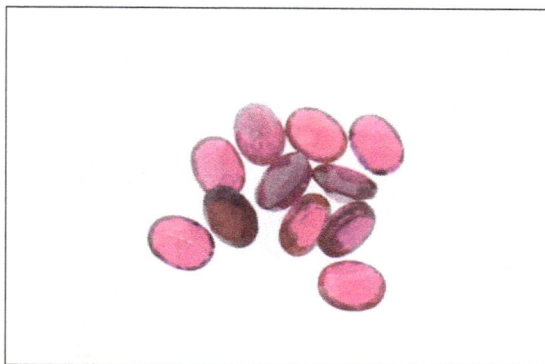

These small organelles perform a number of functions regarding the digestion of compounds such as fats, amino acids, and sugars. They also produce hydrogen peroxide and convert it to water.

Lysosomes

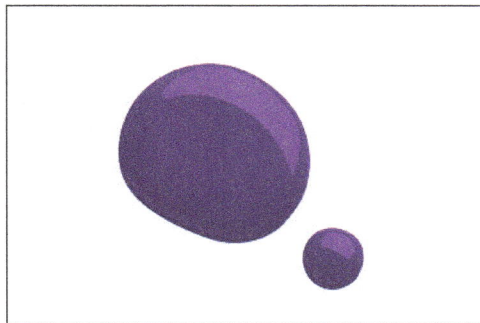

A lysosome is the waste disposal unit of the cell. They are another small organelle and contain a range of enzymes that allow them to digest molecules such as lipids, carbohydrates, and proteins.

Centrosomes

Centrosomes are involved in cell division and the production of flagella and cilia. They consist of two centrioles that are the main hub for a cell's microtubules. As the nuclear envelope breaks down during cell division, microtubules interact with the cell's chromosomes and prepares them for cellular division.

Villi

Villi are needle-like growths that extend from the plasma membrane of a cell. For some cells, such as the cells along the wall of intestines, it is important to be able to rapidly exchange substances with their surrounding environment. Villi increase the rate of exchange of materials between cells and their environment by increasing the surface area of the plasma membrane. This increases the space available for material to move in and out of the cell.

Flagella

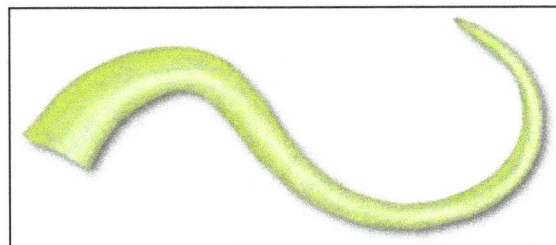

Movement is particularly important for certain animal cells. Sperm cells, for example, live for the sole purpose of traveling to an egg and fertilizing it. Flagella (plural of flagellum) provide the mechanical ability for cells to move under their own power. A flagellum is a long, thin extension of the plasma membrane and is driven by a cellular engine made from proteins.

Different Types of Animal Cells

There are heaps of different types of animal cells and these are just a few from common tissues like skin, muscle, and blood.

Skin Cells

The skin cells of animals mostly consist of keratinocytes and melanocytes – 'cyte' meaning cell. Keratinocytes make up around 90% of all skin cells and produce a protein called 'keratin'. The keratin in skin cells helps to make skin an effective layer of protection for the body. Keratin also makes hair and nails.

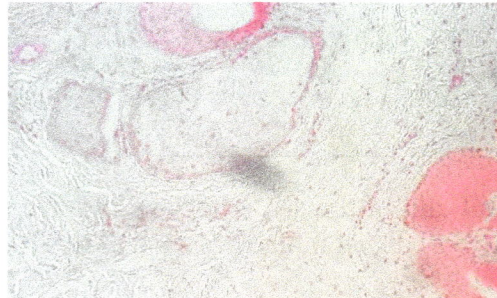

Melanocytes are the second main type of skin cell. They produce a compound called 'melanin' which gives skin its color. Melanocytes sit underneath keratinocytes in a lower layer of skin cells and the melanin they produce is transported up to the surface layers of cells. The more melanocytes you have in your skin, the darker your skin is.

Muscle Cells

Myocytes, muscle fibers or muscle cells are long tubular cells responsible for moving an organism's limbs and organs. Muscle cells can be either skeletal muscle cells, cardiac muscle cells or smooth muscle cells.

Skeletal muscle cells are the most common type of muscle cells and are responsible for making general, conscious movements of the body. Cardiac muscle cells control contractions of heart by generating electrical impulses and smooth muscle cells control subconscious movements of tissues such as blood vessels, the uterus, and the stomach.

Blood Cells

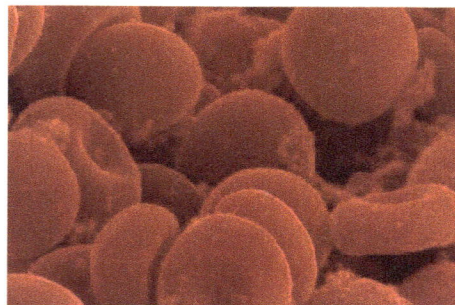

Blood cells can be split into red and white blood cells. Red blood cells make up around 99.9% of all blood cells and are responsible for delivering oxygen from the lungs to the rest of the body. Red blood cells are the only animal cells that do not have a nucleus. White blood cells are a vital part of an animal's immune system and help to battle infections by killing off damaging bacteria and other compounds.

Nerve Cells

Nerve cells, also called neurons, are the main cells of the nervous system. The human brain alone has around 100 billion nerve cells. They are the message carriers of animal cells and deliver and receive signals using dendrites and axons. Dendrites and axons are extensions from the cell that receive and export signals to and from the cell, respectively.

Fat Cells

Fat cells, also known as adipocytes or lipocytes, are used to store fats and other lipids as energy reserves. There are two common types of fat cells in animals – white fat cells and brown fat cells. The main difference between the two cell types is the way they store lipids. White fat cells have one large lipid drop whereas in brown fat cells there are multiple, smaller lipid droplets spread through the cell.

Prokaryotic Cells

Prokaryotes are single-celled organisms that are the earliest and most primitive forms of life on earth. As organized in the Three Domain System, prokaryotes include bacteria and archaeans. Some prokaryotes, such as cyanobacteria, are photosynthetic organisms and are capable of photosynthesis.

Many prokaryotes are extremophiles and are able to live and thrive in various types of extreme environments including hydrothermal vents, hot springs, swamps, wetlands, and the guts of humans and animals (Helicobacter pylori). Prokaryotic bacteria can be found almost anywhere and are part of the human microbiota. They live on your skin, in your body, and on everyday objects in your environment.

Prokaryotic Cell Structure

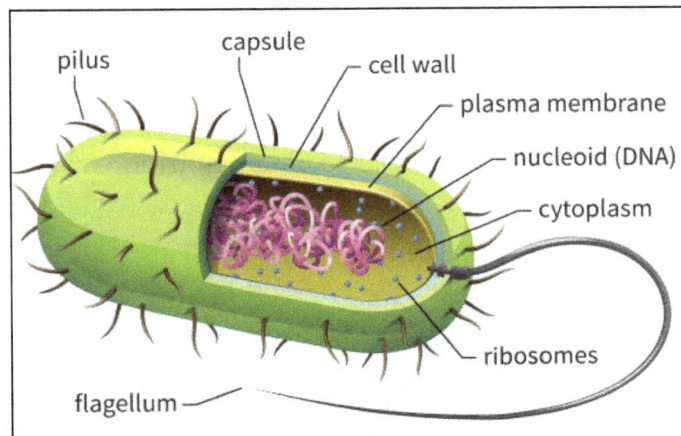

Bacterial Cell Anatomy and Internal Structure.

Prokaryotic cells are not as complex as eukaryotic cells. They have no true nucleus as the DNA is not contained within a membrane or separated from the rest of the cell, but is coiled up in a region of the cytoplasm called the nucleoid. Prokaryotic organisms have varying cell shapes. The most common bacteria shapes are spherical, rod-shaped, and spiral.

Using bacteria as our sample prokaryote, the following structures and organelles can be found in bacterial cells:

- Capsule - Found in some bacterial cells, this additional outer covering protects the cell when it is engulfed by other organisms, assists in retaining moisture, and helps the cell adhere to surfaces and nutrients.

- Cell Wall - The cell wall is an outer covering that protects the bacterial cell and gives it shape.

- Cytoplasm - Cytoplasm is a gel-like substance composed mainly of water that also contains enzymes, salts, cell components, and various organic molecules.

- Cell Membrane or Plasma Membrane - The cell membrane surrounds the cell's cytoplasm and regulates the flow of substances in and out of the cell.

- Pili (Pilus singular)- Hair-like structures on the surface of the cell that attach to other bacterial cells. Shorter pili called fimbriae help bacteria attach to surfaces.

- Flagella - Flagella are long, whip-like protrusion that aids in cellular locomotion.

- Ribosomes - Ribosomes are cell structures responsible for protein production.

- Plasmids - Plasmids are gene carrying, circular DNA structures that are not involved in reproduction.

- Nucleiod Region - Area of the cytoplasm that contains the single bacterial DNA molecule.

Prokaryotic cells lack organelles found in eukaryoitic cells such as mitochondria, endoplasmic reticuli, and Golgi complexes. According to the Endosymbiotic Theory, eukaryotic organelles are thought to have evolved from prokaryotic cells living in endosymbiotic relationships with one another.

Like plant cells, bacteria have a cell wall. Some bacteria also have a polysaccharide capsule layer surrounding the cell wall. It is in this layer where bacteria produce biofilm, a slimy substance that helps bacterial colonies adhere to surfaces and to each other for protection against antibiotics, chemicals, and other hazardous substances.

Similar to plants and algae, some prokaryotes also have photosynthetic pigments. These light-absorbing pigments enable photosynthetic bacteria to obtain nutrition from light.

Binary Fission

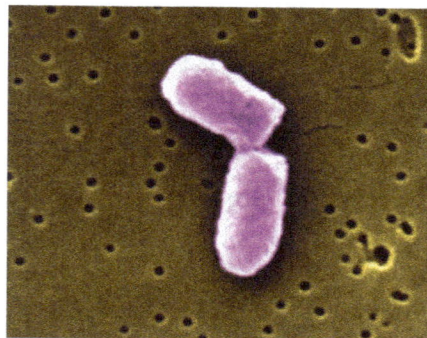

E. coli bacteria undergoing binary fission. The cell wall is dividing resulting in the formation of two cells.

Most prokaryotes reproduce asexually through a process called binary fission. During binary fission, the single DNA molecule replicates and the original cell is divided into two identical cells.

Steps of Binary Fission

- Binary fission begins with DNA replication of the single DNA molecule. Both copies of DNA attach to the cell membrane.

- Next, the cell membrane begins to grow between the two DNA molecules. Once the bacterium just about doubles its original size, the cell membrane begins to pinch inward.

- A cell wall then forms between the two DNA molecules dividing the original cell into two identical daughter cells.

Although E.coli and other bacteria most commonly reproduce by binary fission, this mode of reproduction does not produce genetic variation within the organism.

Prokaryotic Recombination

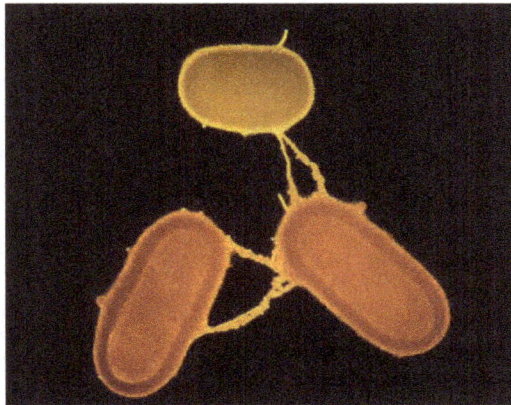

False-color transmission electron micrograph (TEM) of an Escherichia coli bacterium (bottom right) conjugating with two other E.coli bacteria. The tubes connecting the bacteria are pili, which are used to transfer genetic material between bacteria.

Genetic variation within prokaryotic organisms is accomplished through recombination. In recombination, genes from one prokaryote are incorporated into the genome of another prokaryote. Recombination is accomplished in bacterial reproduction by the processes of conjugation, transformation, or transduction.

- In conjugation: Bacteria connect with one another through a protein tube structure called a pilus. Genes are transferred between bacteria through the pilus.

- In transformation: Bacteria take up DNA from their surrounding environment. The DNA is transported across the bacterial cell membrane and incorporated into the bacterial cell's DNA.

- Transduction: Involves the exchange of bacterial DNA through viral infection. Bacteriophages, viruses that infect bacteria, transfer bacterial DNA from previously infected bacteria to any additional bacteria that they infect.

Eukaryotic Cells

Eukaryote is any cell or organism that possesses a clearly defined nucleus. The eukaryotic cell has a nuclear membrane that surrounds the nucleus, in which the well-defined chromosomes (bodies containing the hereditary material) are located. Eukaryotic cells also contain organelles, including mitochondria (cellular energy exchangers), a Golgi apparatus (secretory device), an endoplasmic reticulum (a canal-like system of membranes within the cell), and lysosomes (digestive apparatus within many cell types). There are several exceptions to this, however; for example, the absence of mitochondria and a nucleus in red blood cells and the lack of mitochondria in the oxymonad Monocercomonoides species.

Eukaryotes are thought to have evolved between about 1.7 billion and 1.9 billion years ago. The earliest known microfossils resembling eukaryotic organisms date to approximately 1.8 billion years ago.

Cell Biology

Cell biology is a branch of biology that studies the structure and function of the cell, which is the basic unit of life. Cell biology is concerned with the physiological properties, metabolic processes, signaling pathways, life cycle, chemical composition and interactions of the cell with their environment. This is done both on a microscopic and molecular level as it encompasses prokaryotic cells and eukaryotic cells. Knowing the components of cells and how cells work is fundamental to all biological sciences; it is also essential for research in bio-medical fields such as cancer, and other diseases. Research in cell biology is closely related to genetics, biochemistry, molecular biology, immunology and cytochemistry.

Cell Structure

There are two fundamental classifications of cells: prokaryotes and eukaryotes. The major difference between the two is the presence or absence of organelles. Other factors such as size, the way in which they reproduce, and the number of cells distinguish them from one another. Eukaryotic cells include animal, plant, fungi, and protozoa cells which all have a nucleus enclosed by a membrane, with various shapes and sizes. Prokaryotic cells, lacking an enclosed nucleus, include bacteria and archaea. Prokaryotic cells are much smaller than eukaryotic cells, making prokaryotic cells the smallest form of life. Cytologists typically focus on eukaryotic cells whereas prokaryotic cells are the focus of microbiologists, but this is not always the case.

Cellular Structures

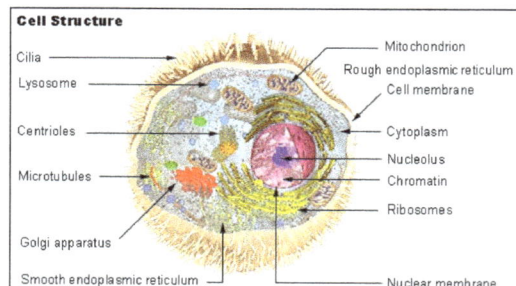

The generalized structure and molecular components of a cell.

Chemical and Molecular Environment

The study of the cell is done on a molecular level; however, most of the processes within the cell are made up of a mixture of small organic molecules, inorganic ions, hormones, and water. Approximately 75-85% of the cell's volume is due to water making it an indispensable solvent as a result of its polarity and structure. These molecules within the cell, which operate as substrates, provide a suitable environment for the cell to carry out metabolic reactions and signalling. The cell shape varies among the different types of organisms, and are thus then classified into two categories: eukaryotes and prokaryotes. In the case of eukaryotic cells - which are made up of animal, plant, fungi, and protozoa cells - the shapes are generally round and spherical or oval while for prokaryotic cells – which are composed of bacteria and archaea - the shapes are: spherical (cocci), rods (bacillus), curved (vibrio), and spirals (*spirochetes*).

Cell biology focuses more on the study of eukaryotic cells, and their signalling pathways, rather than on prokaryotes which is covered under microbiology. The main constituents of the general molecular composition of the cell includes: proteins and lipids which are either free flowing or membrane bound, along with different internal compartments known as organelles. This environment of the cell is made up of hydrophilic and hydrophobic regions which allows for the exchange of the above-mentioned molecules and ions. The hydrophilic regions of the cell are mainly on the inside and outside of the cell, while the hydrophobic regions are within the phospholipid bilayer of the cell membrane. The cell membrane consists of lipids and proteins which accounts for its hydrophobicity as a result of being non-polar substances. Therefore, in order for these molecules to participate in reactions, within the cell, they need to be able to cross this membrane layer to get into the cell. They accomplish this process of gaining access to the cell via: osmotic pressure, diffusion, concentration gradients, and membrane channels. Inside of the cell are extensive internal sub-cellular membrane-bounded compartments called organelles.

Organelles

Cells contain specialized sub-cellular compartments including cell membrane, cytoplasm, mitochondria, and ribosomes.

Processes

Growth and Development

The growth process of the cell does not refer to the size of the cell, but the density of the number of cells present in the organism at a given time. Cell growth pertains to the increase in the number of cells present in an organism as it grows and develops; as the organism gets larger so too does the number of cells present. Cells are the foundation of all organisms, they are the fundamental unit of life. The growth and development of the cell are essential for the maintenance of the host, and survival of the organisms. For this process the cell goes through the steps of the cell cycle and development which involves cell growth, DNA replication, cell division, regeneration, and cell death. The cell cycle is divided into four distinct phases: G1, S, G2, and M. The G phase – which is the cell growth phase – makes up approximately 95% of the cycle. The proliferation of cells is instigated by progenitors; the cells then differentiate to become, where cells of the same type aggregate to form tissues, then organs and ultimately systems. The G phases along with the S phase – DNA replication, damage and repair – are considered to be the inter-phase portion of the cycle, while the M phase (mitosis and [[]]) is the cell division portion of the cycle. The cell cycle is regulated by a series of signalling factors and complexes such as cyclin-dependent kinase and p53, to name a few. When the cell has completed its growth process, and if it is found to be damaged or altered, it undergoes cell death, either by apoptosis or necrosis, in order to eliminate the threat it can cause to the organism's survival.

- Active transport and Passive transport: Movement of molecules into and out of cells.

- Autophagy: The process whereby cells "eat" their own internal components or microbial invaders.

- Adhesion: Holding together cells and tissues.

- Cell movement: Chemotaxis, contraction, cilia and flagella.

- Cell signaling: Regulation of cell behavior by signals from outside.

Other Cellular Processes

- Division: By which cells reproduce either by mitosis (to produce clones of the parent cell) or Meiosis (to produce haploid gametes).

- DNA repair: Cell death and cell senescence.

- Metabolism: Glycolysis, respiration, photosynthesis, and chemosynthesis.

- Signalling: The process by which the activities in the cell are regulated.

- Transcription and mRNA splicing - Gene expression.

Active Transport

In cellular biology, active transport is the movement of molecules across a membrane from a region of their lower concentration to a region of their higher concentration—against the concentration gradient. Active transport requires cellular energy to achieve this movement. There are two types of active transport: primary active transport that uses adenosine triphosphate (ATP), and secondary active transport that uses an electrochemical gradient. An example of active transport in human physiology is the uptake of glucose in the intestines.

Active Cellular Transportation

Unlike passive transport, which uses the kinetic energy and natural entropy of molecules moving down a gradient, active transport uses cellular energy to move them against a gradient, polar repulsion, or other resistance. Active transport is usually associated with accumulating high concentrations of molecules that the cell needs, such as ions, glucose and amino acids. If the process uses chemical energy, such as from adenosine triphosphate (ATP), it is termed primary active transport. Secondary active transport involves the use of an electrochemical gradient. Examples of active transport include the uptake of glucose in the intestines in humans and the uptake of mineral ions into root hair cells of plants.

Primary Active Transport

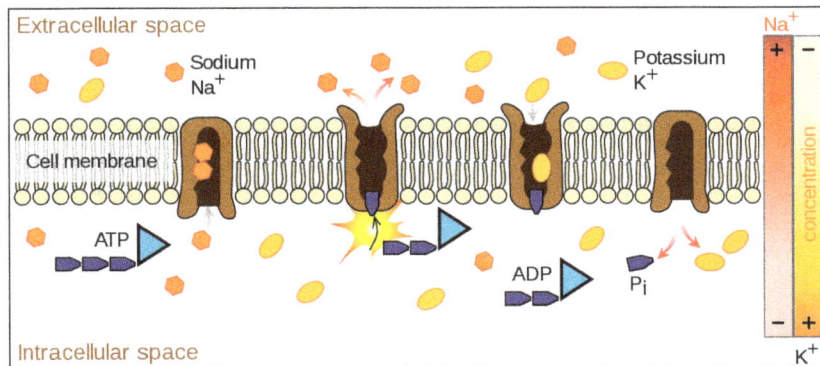

The action of the sodium-potassium pump is an example of primary active transport.

Primary active transport, also called direct active transport, directly uses metabolic energy to transport molecules across a membrane. Substances that are transported across the cell membrane by primary active transport include metal ions, such as Na^+, K^+, Mg^{2+}, and Ca^{2+}. These charged particles require ion pumps or ion channels to cross membranes and distribute through the body.

Most of the enzymes that perform this type of transport are transmembrane ATPases. A primary ATPase universal to all animal life is the sodium-potassium pump, which helps to maintain the cell potential. The sodium-potassium pump maintains the membrane potential by moving three Na+ ions out of the cell for every two K+ ions moved into the cell. Other sources of energy for Primary active transport are redox energy and photon energy (light). An example of primary active transport using Redox energy is the mitochondrial electron transport chain that uses the reduction energy of NADH to move protons across the inner mitochondrial membrane against their concentration gradient. An example of primary active transport using light energy are the proteins involved in photosynthesis that use the energy of photons to create a proton gradient across the thylakoid membrane and also to create reduction power in the form of NADPH.

Model of Active Transport

ATP hydrolysis is used to transport hydrogen ions against the electrochemical gradient (from low to high hydrogen ion concentration). Phosphorylation of the carrier protein and the binding of a hydrogen ion induce a conformational (shape) change that drives the hydrogen ions to transport against the electrochemical gradient. Hydrolysis of the bound phosphate group and release of hydrogen ion then restores the carrier to its original conformation.

Types of Primary Active Transporters

1. P-type ATPase: Sodium potassium pump, calcium pump, proton pump.

2. F-ATPase: Mitochondrial ATP synthase, chloroplast ATP synthase.

3. V-ATPase: Vacuolar ATPase.

4. ABC (ATP binding cassette) transporter: MDR, CFTR, etc.

Adenosine Triphosphate-binding cassette transporters (ABC transporters) comprise a large and diverse protein family, often functioning as ATP-driven pumps. Usually, there are several domains involved in the overall transporter protein's structure, including two nucleotide-binding domains that constitute the ATP-binding motif and two hydrophobic transmembrane domains that create the "pore" component. In broad terms, ABC transporters are involved in the import or export of molecules across a cell membrane; yet within the protein family there is an extensive range of function.

In plants, ABC transporters are often found within cell and organelle membranes, such as the mitochondria, chloroplast, and plasma membrane. There is evidence to support that plant ABC transporters play a direct role in pathogen response, phytohormone transport, and detoxification. Furthermore, certain plant ABC transporters may function in actively exporting volatile compounds and antimicrobial metabolites.

In petunia flowers (*Petunia hybrida*), the ABC transporter PhABCG1 is involved in the active transport of volatile organic compounds. PhABCG1 is expressed in the petals of open flowers. In general, volatile compounds may promote the attraction of seed-dispersal organisms and pollinators, as well as aid in defense, signaling, allelopathy, and protection. To study the protein PhABCG1, transgenic petunia RNA interference lines were created with decreased *PhABCG1* expression levels. In these transgenic lines, a decrease in emission of volatile compounds was observed. Thus, PhABCG1 is likely involved in the export of volatile compounds. Subsequent experiments involved incubating control and transgenic lines that expressed *PhABCG1* to test for transport activity involving different substrates. Ultimately, PhABCG1 is responsible for the protein-mediated transport of volatile organic compounds, such as benezyl alcohol and methylbenzoate, across the plasma membrane.

Additionally in plants, ABC transporters may be involved in the transport of cellular metabolites. Pleiotropic Drug Resistance ABC transporters are hypothesized to be involved in stress response and export antimicrobial metabolites. One example of this type of ABC transporter is the protein NtPDR1. This unique ABC transporter is found in *Nicotiana tabacum* BY2 cells and is expressed in the presence of microbial elicitors. NtPDR1 is localized in the root epidermis and aerial trichomes of the plant. Experiments using antibodies specifically targeting NtPDR1 followed by Western blotting allowed for this determination of localization. Furthermore, it is likely that the protein NtPDR1 actively transports out antimicrobial diterpene molecules, which are toxic to the cell at high levels.

Secondary Active Transport

In secondary active transport, also known as *coupled transport* or *cotransport*, energy is used to transport molecules across a membrane; however, in contrast to primary active transport, there is no direct coupling of ATP; instead it relies upon the electrochemical potential difference created by pumping ions in/out of the cell. Permitting one ion or molecule to move down an electrochemical gradient, but possibly against the concentration gradient where it is more concentrated to that where it is less concentrated increases entropy and can serve as a source of energy for metabolism (e.g. in ATP synthase). The energy derived from the pumping of protons across a cell membrane is frequently used as the energy source in secondary active transport. In humans, sodium (Na^+) is a commonly co-transported ion across the plasma membrane, whose electrochemical gradient is then used to power the active transport of a second ion or molecule against its gradient. In bacteria and small yeast cells, a commonly cotransported ion is hydrogen. Hydrogen pumps are also used to create an electrochemical gradient to carry out processes within cells such as in the electron transport chain, an important function of cellular respiration that happens in the mitochondrion of the cell.

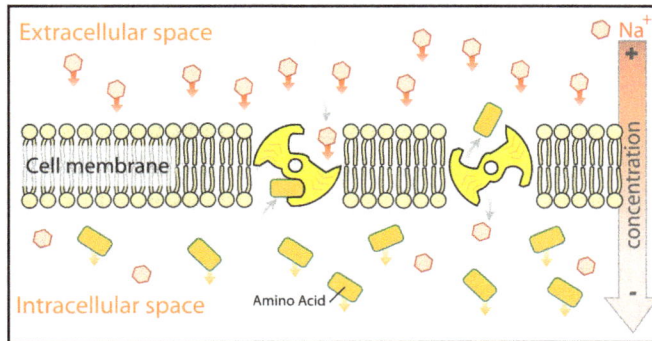

Secondary active transport

In August 1960, in Prague, Robert K. Crane presented for the first time his discovery of the sodium-glucose cotransport as the mechanism for intestinal glucose absorption. Crane's discovery of cotransport was the first ever proposal of flux coupling in biology.

Cotransporters can be classified as symporters and antiporters depending on whether the substances move in the same or opposite directions.

Antiporter

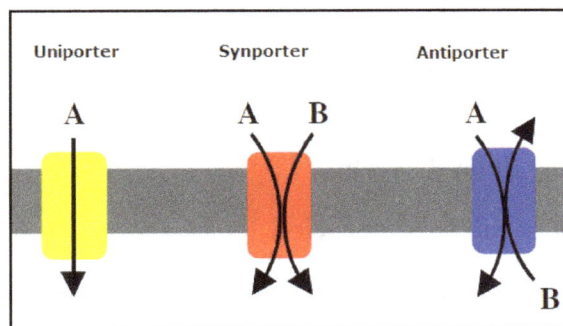

Function of symporters and antiporters.

In an antiporter two species of ion or other solutes are pumped in opposite directions across a membrane. One of these species is allowed to flow from high to low concentration which yields the entropic energy to drive the transport of the other solute from a low concentration region to a high one.

An example is the sodium-calcium exchanger or antiporter, which allows three sodium ions into the cell to transport one calcium out. This antiporter mechanism is important within the membranes of cardiac muscle cells in order to keep the calcium concentration in the cytoplasm low. Many cells also possess calcium ATPases, which can operate at lower intracellular concentrations of calcium and sets the normal or resting concentration of this important second messenger. But the ATPase exports calcium ions more slowly: only 30 per second versus 2000 per second by the exchanger. The exchanger comes into service when the calcium concentration rises steeply or "spikes" and enables rapid recovery. This shows that a single type of ion can be transported by several enzymes, which need not be active all the time (constitutively), but may exist to meet specific, intermittent needs.

Symporter

A symporter uses the downhill movement of one solute species from high to low concentration to move another molecule uphill from low concentration to high concentration (against its concentration gradient). Both molecules are transported in the same direction.

An example is the glucose symporter SGLT1, which co-transports one glucose (or galactose) molecule into the cell for every two sodium ions it imports into the cell. This symporter is located in the small intestines, heart, and brain. It is also located in the S3 segment of the proximal tubule in each nephron in the kidneys. Its mechanism is exploited in glucose rehydration therapy. This mechanism uses the absorption of sugar through the walls of the intestine to pull water in along with it. Defects in SGLT2 prevent effective reabsorption of glucose, causing familial renal glucosuria.

Bulk Transport

Endocytosis and exocytosis are both forms of bulk transport that move materials into and out of cells, respectively, via vesicles. In the case of endocytosis, the cellular membrane folds around the desired materials outside the cell. The ingested particle becomes trapped within a pouch, known as a vesicle, inside the cytoplasm. Often enzymes from lysosomes are then used to digest the molecules absorbed by this process. Substances that enter the cell via signal mediated electrolysis include proteins, hormones and growth and stabilization factors. Viruses enter cells through a form of endocytosis that involves their outer membrane fusing with the membrane of the cell. This forces the viral DNA into the host cell.

Biologists distinguish two main types of endocytosis: pinocytosis and phagocytosis.

- In pinocytosis, cells engulf liquid particles (in humans this process occurs in the small intestine, where cells engulf fat droplets).

- In phagocytosis, cells engulf solid particles.

Exocytosis involves the removal of substances through the fusion of the outer cell membrane and a vesicle membrane. An example of exocytosis would be the transmission of neurotransmitters across a synapse between brain cells.

Passive Transport

Passive transport is a movement of ions and other atomic or molecular substances across cell membranes without need of energy input. Unlike active transport, it does not require an input of cellular energy because it is instead driven by the tendency of the system to grow in entropy. The rate of passive transport depends on the permeability of the cell membrane, which, in turn, depends on the organization and characteristics of the membrane lipids and proteins. The four main kinds of passive transport are simple diffusion, facilitated diffusion, filtration, and/or osmosis.

Diffusion

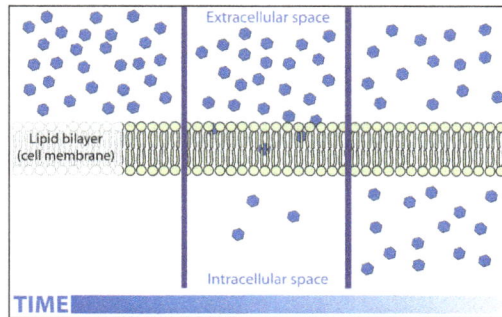

Passive diffusion on a cell membrane.

Diffusion is the net movement of material from an area of high concentration to an area with lower concentration. The difference of concentration between the two areas is often termed as the *concentration gradient*, and diffusion will continue until this gradient has been eliminated. Since diffusion moves materials from an area of higher concentration to an area of lower concentration, it is described as moving solutes "down the concentration gradient" (compared with active transport, which often moves material from area of low concentration to area of higher concentration, and therefore referred to as moving the material "against the concentration gradient"). However, in many cases (e.g. passive drug transport) the driving force of passive transport can not be simplified to the concentration gradient. If there are different solutions at the two sides of the membrane with different equilibrium solubility of the drug, the difference in degree of saturation is the driving force of passive membrane transport. It is also true for supersaturated solutions which are more and more important owing to the spreading of the application of amorphous solid dispersions for drug bioavailability enhancement.

Simple diffusion and osmosis are in some ways similar. Simple diffusion is the passive movement of solute from a high concentration to a lower concentration until the concentration of the solute is uniform throughout and reaches equilibrium. Osmosis is much like simple diffusion but it specifically describes the movement of water (not the solute) across a selectively permeable membrane until there is an equal concentration of water and solute on both sides of the membrane. Simple diffusion and osmosis are both forms of passive transport and require none of the cell's [Adenosine triphosphate [ATP] energy].

Facilitated Diffusion

Facilitated diffusion, also called carrier-mediated osmosis, is the movement of molecules across the cell membrane via special transport proteins that are embedded in the plasma membrane by actively taking up or excluding ions. Active transport of protons by H^+ ATPases alters membrane potential allowing for facilitated passive transport of particular ions such as potassium down their charge gradient through high affinity transporters and channels.

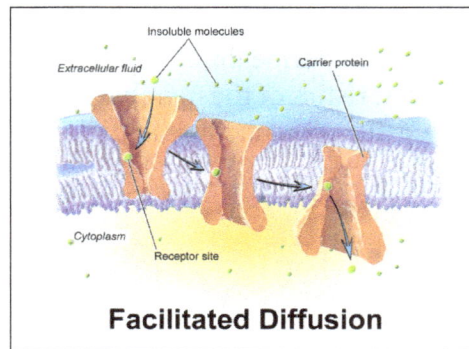

Facilitated Diffusion

Depiction of facilitated diffusion.

Filtration

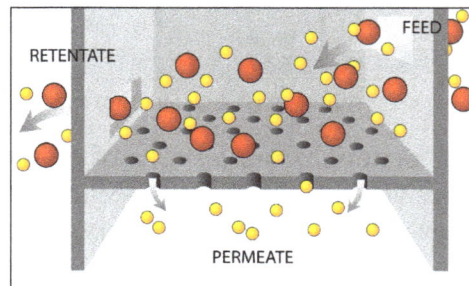

Filtration.

Filtration is movement of water and solute molecules across the cell membrane due to hydrostatic pressure generated by the cardiovascular system. Depending on the size of the membrane pores, only solutes of a certain size may pass through it. For example, the membrane pores of the Bowman's capsule in the kidneys are very small, and only albumins, the smallest of the proteins, have any chance of being filtered through. On the other hand, the membrane pores of liver cells are extremely large, but not forgetting cells are extremely small to allow a variety of solutes to pass through and be metabolized.

Osmosis

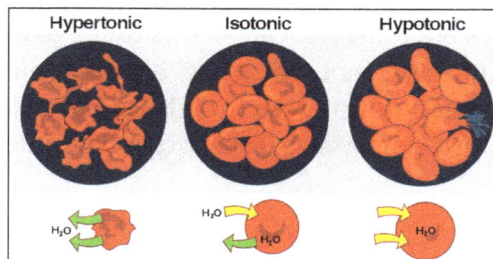

Effect of osmosis on blood cells under different solutions.

Osmosis is the movement of water molecules across a selectively permeable membrane. The net movement of water molecules through a partially permeable membrane from a solution of high water potential to an area of low water potential. A cell with a less negative water potential will draw in water but this depends on other factors as well such as solute potential (pressure in the cell e.g. solute molecules) and pressure potential (external pressure e.g. cell wall). There are three types of Osmosis solutions: the isotonic solution, hypotonic solution, and hypertonic solution. Isotonic solution is when the extracellular solute concentration is balanced with the concentration inside the cell. In the Isotonic solution, the water molecules still moves between the solutions, but the rates are the same from both directions, thus the water movement is balanced between the inside of the cell as well as the outside of the cell. A hypotonic solution is when the solute concentration outside the cell is lower than the concentration inside the cell. In hypotonic solutions, the water moves into the cell, down its concentration gradient (from higher to lower water concentrations). That can cause the cell to swell. Cells that don't have a cell wall, such as animal cells, could burst in this solution. A hypertonic solution is when the solute concentration is higher (think of hyper - as high) than the concentration inside the cell. In hypertonic solution, the water will move out, causing the cell to shrink.

Autophagy

Autophagy (or *autophagocytosis*) is the natural, regulated mechanism of the cell that removes unnecessary or dysfunctional components. It allows the orderly degradation and recycling of cellular components.

Three forms of autophagy are commonly described: macroautophagy, microautophagy, and chaperone-mediated autophagy (CMA). In macroautophagy, expendable cytoplasmic constituents are targeted and isolated from the rest of the cell within a double-membraned vesicle known as an autophagosome, which, in time, fuses with an available lysosome, bringing its specialty process of waste management and disposal; and eventually the contents of the vesicle (now called an autolysosome) are degraded and recycled.

In disease, autophagy has been seen as an adaptive response to stress, promoting survival of the cell; but in other cases it appears to promote cell death and morbidity. In the extreme case of starvation, the breakdown of cellular components promotes cellular survival by maintaining cellular energy levels.

The name "autophagy" was in existence and frequently used from the middle of the 19th century. In its present usage, the term autophagy was coined by Belgian biochemist Christian de Duve in 1963 based on his discovery of the functions of lysosome. The identification of autophagy-related genes in yeast in the 1990s allowed researchers to deduce the mechanisms of autophagy, which eventually led to the award of the 2016 Nobel Prize in Physiology or Medicine to Japanese researcher Yoshinori Ohsumi.

Autophagy was first observed by Keith R. Porter and his student Thomas Ashford at the Rockefeller Institute. In January 1962 they reported an increased number of lysosomes in rat liver cells after the addition of glucagon, and that some displaced lysosomes towards the centre of the cell contained other cell organelles such as mitochondria. They called this autolysis after Christian de Duve and Alex B. Novikoff. However Porter and Ashford wrongly interpreted their data as lysosome formation (ignoring the pre-existing organelles). Lysosomes could not be cell organelles, but part of cytoplasm such as mitochondria, and that hydrolytic enzymes were produced by microbodies. In 1963 Hruban, Spargo and colleagues published a detailed ultrastructural description of "focal cytoplasmic degradation," which referenced a 1955 German study of injury-induced sequestration. Hruban, Spargo and colleagues recognized three continuous stages of maturation of the sequestered cytoplasm to lysosomes, and that the process was not limited to injury states that functioned under physiological conditions for "reutilization of cellular materials," and the "disposal of organelles" during differentiation. Inspired by this discovery, de Duve christened the phenomena "autophagy". Unlike Porter and Ashford, de Duve conceived the term as a part of lysosomal function while describing the role of glucagon as a major inducer of cell degradation in the liver. With his student Russell Deter, he established that lysosomes are responsible for glucagon-induced autophagy. This was the first time the fact that lysosomes are the sites of intracellular autophagy was established.

In the 1990s, several groups of scientists independently discovered autophagy-related genes using the budding yeast. Notably, Yoshinori Ohsumi and Michael Thumm examined starvation-induced non-selective autophagy; in the meantime, Daniel J Klionsky discovered the cytoplasm-to-vacuole targeting (CVT) pathway, which is a form of selective autophagy. They soon found that they were in fact looking at essentially the same pathway, just from different angles. Initially, the genes discovered by these and other yeast groups were given different names (APG, AUT, CVT, GSA, PAG, PAZ, and PDD). A unified nomenclature was advocated in 2003 by the yeast researchers to use ATG to denote autophagy genes. The 2016 Nobel Prize in Physiology or Medicine was awarded to Yoshinori Ohsumi, although some have pointed out that the award could have been more inclusive.

The field of autophagy research experienced accelerated growth at the turn of the 21st century. Knowledge of ATG genes provided scientists more convenient tools to dissect functions of autophagy in human health and disease. In 1999, a landmark discovery connecting autophagy with cancer was published by Beth Levine's group. To this date, relationship between cancer and autophagy continues to be a main theme of autophagy research. The roles of autophagy in neurodegeneration and immune defense also received considerable attention. In 2003, the first Gordon Research Conference on autophagy was held at Waterville. In 2005, Daniel J Klionsky launched *Autophagy*, a scientific journal dedicated to this field. The first Keystone Symposia Conference on autophagy was held in 2007 at Monterey. In 2008, Carol A Mercer created a BHMT fusion protein (GST-BHMT), which showed starvation-induced site-specific fragmentation in cell lines. The degradation of betaine homo-cysteine methyltransferase (BHMT), a metabolic enzyme, could be used to assess autophagy flux in mammalian cells.

In contemporary literature, the brazilian writer Leonid Bózio expresses autophagy as an existential question. The psychological drama of the book *Tempos Sombrios* recounts characters consuming their own lives in an inauthentic existence.

Process and Pathways

There are three main types of autophagy, namely macroautophagy, microautophagy and Chaperone mediated autophagy. They are mediated by the autophagy-related genes and their associated enzymes. Macroautophagy is then divided into bulk and selective autophagy. In the selective autophagy is the autophagy of organelles; mitophagy, lipophagy, pexophagy, chlorophagy, ribophagy and others.

Macroautophagy is the main pathway, used primarily to eradicate damaged cell organelles or unused proteins. First the phagophore engulfs the material that needs to be degraded, which forms a double membrane known as an autophagosome, around the organelle marked for destruction. The autophagosome then travels through the cytoplasm of the cell to a lysosome, and the two organelles fuse. Within the lysosome, the contents of the autophagosome are degraded via acidic lysosomal hydrolase.

Microautophagy, on the other hand, involves the direct engulfment of cytoplasmic material into the lysosome. This occurs by invagination, meaning the inward folding of the lysosomal membrane, or cellular protrusion.

Chaperone-mediated autophagy, or CMA, is a very complex and specific pathway, which involves the recognition by the hsc70-containing complex. This means that a protein must contain the recognition site for this hsc70 complex which will allow it to bind to this chaperone, forming the CMA- substrate/chaperone complex. This complex then moves to the lysosomal membrane-bound protein that will recognise and bind with the CMA receptor, allowing it to enter the cell. Upon recognition, the substrate protein gets unfolded and it is translocated across the lysosome membrane with the assistance of the lysosomal hsc70 chaperone. CMA is significantly different from other types of autophagy because it translocates protein material in a one by one manner, and it is extremely selective about what material crosses the lysosomal barrier.

Mitophagy is the selective degradation of mitochondria by autophagy. It often occurs to defective mitochondria following damage or stress. Mitophagy promotes turnover of mitochondria and prevents accumulation of dysfunctional mitochondria which can lead to cellular degeneration. It is mediated by Atg32 (in yeast) and NIX and its regulator BNIP3 in mammals. Mitophagy is regulated by PINK1 and parkin proteins. The occurrence of mitophagy is not limited to the damaged mitochondria but also involves undamaged ones.

Lipophagy is the degradation of lipids by autophagy, a function which has been shown to exist in both animal and fungal cells. The role of lipophagy in plant cells, however, remains elusive. In lipophagy the target are lipid structures called lipid droplets (LDs), spheric "organelles" with a core of mainly triacylglycerols (TAGs) and a unilayer of phospholipids and membrane proteins. In animal cells the main lipophagic pathway is via the engulfment of LDs by the phagophore, macroautophagy. In fungal cells on the other hand microplipophagy constitutes the main pathway and is especially well studied in the budding yeast *Saccharomyces cerevisiae*. Lipophagy was first discovered in mice and published 2009.

Molecular Biology

Autophagy is executed by autophagy-related (Atg) genes. Prior to 2003, ten or more names were used, but after this point a unified nomenclature was devised by fungal autophagy researchers. Atg or ATG stands for autophagy related. It does not specify gene or a protein.

The first autophagy genes were identified by genetic screens conducted in *Saccharomyces cerevisiae*. Following their identification those genes were functionally characterized and their orthologs in a variety of different organisms were identified and studied.

In mammals, amino acid sensing and additional signals such as growth factors and reactive oxygen species regulate the activity of the protein kinases mTOR and AMPK. These two kinases regulate autophagy through inhibitory phosphorylation of the Unc-51-like kinases ULK1 and ULK2 (mammalian homologues of Atg1). Induction of autophagy results in the dephosphorylation and activation of the ULK kinases. ULK is part of a protein complex containing Atg13, Atg101 and FIP200. ULK phosphorylates and activates Beclin-1 (mammalian homologue of Atg6), which is also part of a protein complex. The autophagy-inducible Beclin-1 complex contains the proteins p150, Atg14L and the class III phosphatidylinositol 3-phosphate kinase (PI(3)K) Vps34. The active ULK and Beclin-1 complexes re-localize to the site of autophagosome initiation, the phagophore, where they both contribute to the activation of downstream autophagy components.

Once active, VPS34 phosphorylates the lipid phosphatidylinositol to generate phosphatidylinositol 3-phosphate (PtdIns(3)P) on the surface of the phagophore. The generated PtdIns(3)P is used as a docking point for proteins harboring a PtdIns(3)P binding motif. WIPI2, a PtdIns(3)P binding protein of the WIPI (WD-repeat protein interacting with phosphoinositides) protein family, was recently shown to physically bind Atg16L1. Atg16L1 is a member of an E3-like protein complex involved in one of two ubiquitin-like conjugation systems essential for autophagosome formation. Its binding by WIPI2 recruits it to the phagophore and mediates its activity.

The first of the two ubiquitin-like conjugation systems involved in autophagy covalently binds the ubiquitin-like protein Atg12 to Atg5. The resulting conjugate protein then binds Atg16L1 to form an E3-like complex which functions as part of the second ubiquitin-like conjugation system. This complex binds and activates Atg3, which covalently attaches mammalian homologues of the ubiquitin-like yeast protein ATG8 (LC3A-C, GATE16, and GABARAPL1-3), the most studied being LC3 proteins, to the lipid phosphatidylethanolamine (PE) on the surface of autophagosomes. Lipidated LC3 contributes to the closure of autophagosomes, and enables the docking of specific cargos and adaptor proteins such as Sequestosome-1/p62. The completed autophagosome then fuses with a lysosome through the actions of multiple proteins, including SNAREs and UVRAG. Following the fusion LC3 is retained on the vesicle's inner side and degraded along with the cargo, while the LC3 molecules attached to the outer side are cleaved off by Atg4 and recycled. The contents of the autolysosome are subsequently degraded and their building blocks are released from the vesicle through the action of permeases.

Functions

Nutrient Starvation

Autophagy has roles in various cellular functions. One particular example is in yeasts, where the nutrient starvation induces a high level of autophagy. This allows unneeded proteins to be degraded and the amino acids recycled for the synthesis of proteins that are essential for survival. In higher eukaryotes, autophagy is induced in response to the nutrient depletion that occurs in animals at birth after severing off the trans-placental food supply, as well as that of nutrient starved cultured cells and tissues. Mutant yeast cells that have a reduced autophagic capability rapidly perish in nutrition-deficient conditions. Studies on the *apg* mutants suggest that autophagy via autophagic bodies is indispensable for protein degradation in the vacuoles under starvation conditions, and that at least 15 APG genes are involved in autophagy in yeast. A gene known as ATG7 has been implicated in nutrient-mediated autophagy, as mice studies have shown that starvation-induced autophagy was impaired in *atg7*-deficient mice.

Xenophagy

In microbiology, xenophagy is the autophagic degradation of infectious particles. Cellular auto-phagic machinery also play an important role in innate immunity. Intracellular pathogens, such as *Mycobacterium tuberculosis* (the bacterium which is responsible for tuberculosis) are targeted for degradation by the same cellular machinery and regulatory mechanisms that target host mito-chondria for degradation. Incidentally, this is further evidence for the endosymbiotic hypothesis. This process generally leads to the destruction of the invasive microorganism, although some bac-teria can block the maturation of phagosomes into degradative organelles called phagolysosomes. Stimulation of autophagy in infected cells can help overcome this phenomenon, restoring patho-gen degradation.

Infection

Vesicular stomatitis virus is believed to be taken up by the autophagosome from the cytosol and translocated to the endosomes where detection takes place by a pattern recognition receptor called toll-like receptor 7, detecting single stranded RNA. Following activation of the toll-like receptor, intracellular signaling cascades are initiated, leading to induction of interferon and other antiviral cytokines. A subset of viruses and bacteria subvert the autophagic pathway to promote their own replication. Galectin-8 has recently been identified as an intracellular "danger receptor", able to initiate autophagy against intracellular pathogens. When galectin-8 binds to a damaged vacuole, it recruits an autophagy adaptor such as NDP52 leading to the formation of an autophagosome and bacterial degradation.

Repair Mechanism

Autophagy degrades damaged organelles, cell membranes and proteins, and electing against au-tophagy is thought to be one of the main reasons for the accumulation of damaged cells and aging. Autophagy and autophagy regulators are involved in response to lysosomal damage, often direct-ed by galectins such as galectin-3 and galectin-8, which in turn recruit receptors such as TRIM16 and NDP52 plus directly affect mTOR and AMPK activity, whereas mTOR and AMPK inhibit and activate autophagy, respectively.

Programmed Cell Death

One of the mechanisms of programmed cell death (PCD) is associated with the appearance of autophagosomes and depends on autophagy proteins. This form of cell death most likely corre-sponds to a process that has been morphologically defined as autophagic PCD. One question that constantly arises, however, is whether autophagic activity in dying cells is the cause of death or is actually an attempt to prevent it. Morphological and histochemical studies so far did not prove a causative relationship between the autophagic process and cell death. In fact, there have re-cently been strong arguments that autophagic activity in dying cells might actually be a survival mechanism. Studies of the metamorphosis of insects have shown cells undergoing a form of PCD that appears distinct from other forms; these have been proposed as examples of autophagic cell death. Recent pharmacological and biochemical studies have proposed that survival and lethal autophagy can be distinguished by the type and degree of regulatory signaling during stress particularly after viral infection. Although promising, these findings have not been examined in non-viral systems.

Exercise

Autophagy is essential for basal homeostasis; it is also extremely important in maintaining muscle homeostasis during physical exercise. Autophagy at the molecular level is only partially understood. A study of mice shows that autophagy is important for the ever-changing demands of their nutritional and energy needs, particularly through the metabolic pathways of protein catabolism. In a 2012 study conducted by the University of Texas Southwestern Medical Center in Dallas, mutant mice (with a knock-in mutation of BCL2 phosphorylation sites to produce progeny that showed normal levels of basal autophagy yet were deficient in stress-induced autophagy) were tested to challenge this theory. Results showed that when compared to a control group, these mice illustrated a decrease in endurance and an altered glucose metabolism during acute exercise.

Another study demonstrated that skeletal muscle fibres of collagen VI knockout mice showed signs of degeneration due to an insufficiency of autophagy which led to an accumulation of damaged mitochondria and excessive cell death. Exercise-induced autophagy was unsuccessful however; but when autophagy was induced artificially post-exercise, the accumulation of damaged organelles in collagen VI deficient muscle fibres was prevented and cellular homeostasis was maintained. Both studies demonstrate that autophagy induction may contribute to the beneficial metabolic effects of exercise and that it is essential in the maintaining of muscle homeostasis during exercise, particularly in collagen VI fibres.

Work at the Institute for Cell Biology, University of Bonn, showed that a certain type of autophagy, i.e. chaperone-assisted selective autophagy (CASA), is induced in contracting muscles and is required for maintaining the muscle sarcomere under mechanical tension. The CASA chaperone complex recognizes mechanically damaged cytoskeleton components and directs these components through a ubiquitin-dependent autophagic sorting pathway to lysosomes for disposal. This is necessary for maintaining muscle activity.

Osteoarthritis

Because autophagy decreases with age and age is a major risk factor for osteoarthritis, the role of autophagy in the development of this disease is suggested. Proteins involved in autophagy are reduced with age in both human and mouse articular cartilage. Mechanical injury to cartilage explants in culture also reduced autophagy proteins. Autophagy is constantly activated in normal cartilage but it is compromised with age and precedes cartilage cell death and structural damage. Thus autophagy is involved in a normal protective process (chondroprotection) in the joint.

Parkinson Disease

Parkinson disease is a neurodegenerative disorder partially caused by the cell death of brain and brain stem cells in many nuclei like the substantia nigra. Parkinson's disease is characterized by inclusions of a protein called alpha-synuclien (Lewy bodies) in affected neurons that cells cannot break down. Deregulation of the autophagy pathway and mutation of alleles regulating autophagy are believed to cause neurodegenerative diseases. Autophagy is essential for neuronal survival.

Without efficient autophagy, neurons gather ubiquitinated protein aggregates and degrade. Ubiquitinated proteins are proteins that have been tagged with ubiquitin to get degraded. Mutations of synuclien alleles lead to lysosome pH increase and hydrolase inhibition. As a result, lysosomes degradative capacity is decreased. There are several genetic mutations implicated in the disease, including loss of function PINK1 and Parkin. Loss of function in these genes can lead to damaged mitochondrial accumulation and protein aggregates than can lead to cellular degeneration. Mitochondria is involved in Parkinson's disease. In idiopathic Parkinson's disease, the disease is commonly caused by dysfunctional mitochondria, cellular oxidative stress, autophagic alterations and the aggregation of proteins. These can lead to mitochondrial swelling and depolarization.

Significance of Autophagy as a Drug Target

Since dysregulation of autophagy is involved in the pathogenesis of a broad range of diseases, great efforts are invested to identify and characterize small synthetic or natural molecules that can regulate it.

Cell Division

Cell division is the process by which a parent cell divides into two or more daughter cells. Cell division usually occurs as part of a larger cell cycle. In eukaryotes, there are two distinct types of cell division: a vegetative division, whereby each daughter cell is genetically identical to the parent cell (mitosis), and a reproductive cell division, whereby the number of chromosomes in the daughter cells is reduced by half to produce haploid gametes (meiosis). Meiosis results in four haploid daughter cells by undergoing one round of DNA replication followed by two divisions. Homologous chromosomes are separated in the first division, and sister chromatids are separated in the second division. Both of these cell division cycles are used in the process of sexual reproduction at some point in their life cycle. Both are believed to be present in the last eukaryotic common ancestor.

Prokaryotes (bacteria) undergo a vegetative cell division known as binary fission, where their genetic material is segregated equally into two daughter cells. While binary fission may be the means of division by most prokaryotes, there are alternative manners of division, such as budding, that have been observed. All cell divisions, regardless of organism, are preceded by a single round of DNA replication.

For simple unicellular microorganisms such as the amoeba, one cell division is equivalent to reproduction – an entire new organism is created. On a larger scale, mitotic cell division can create progeny from multicellular organisms, such as plants that grow from cuttings. Mitotic cell division enables sexually reproducing organisms to develop from the one-celled zygote, which itself was produced by meiotic cell division from gametes. After growth, cell division by mitosis allows for continual construction and repair of the organism. The human body experiences about 10 quadrillion cell divisions in a lifetime.

The primary concern of cell division is the maintenance of the original cell's genome. Before division can occur, the genomic information that is stored in chromosomes must be replicated, and the duplicated genome must be separated cleanly between cells. A great deal of cellular infrastructure is involved in keeping genomic information consistent between generations.

Phases of Eukaryotic Cell Division

Interphase

Interphase is the process a cell must go through before mitosis, meiosis, and cytokinesis. Interphase consists of three main phases: G_1, S, and G_2. G_1 is a time of growth for the cell where specialized cellular functions occur in order to prepare the cell for DNA Replication. There are checkpoints during interphase that allow the cell to be either advance or halt further development. In S phase, the chromosomes are replicated in order for the genetic content to be maintained. During G_2, the cell undergoes the final stages of growth before it enters the M phase, where spindles are synthesized. The M phase, can be either mitosis or meiosis depending on the type of cell. Germ cells, or gametes, undergo meiosis, while somatic cells will undergo mitosis. After the cell proceeds successfully through the M phase, it may then undergo cell division through cytokinesis. The control of each checkpoint is controlled by cyclin and cyclin-dependent kinases. The progression of interphase is the result of the increased amount of cyclin. As the amount of cyclin increases, more and more cyclin dependent kinases attach to cyclin signaling the cell further into interphase. At the peak of the cyclin attached to the cyclin dependent kinases this system pushes the cell out of interphase and into the M phase, where mitosis, meiosis, and cytokinesis occur. There are three transition checkpoints the cell goes through before entering the M phase. The most important being the G_1-S transition checkpoint. If the cell does not pass this checkpoint, then the cell will exit the cell cycle.

Prophase

Prophase is the first stage of division. The nuclear envelope is broken down, long strands of chromatin condense to form shorter more visible strands called chromosomes, the nucleolus disappears, and microtubules attach to the chromosomes at the kinetochores present in the centromere. Microtubules associated with the alignment and separation of chromosomes are referred to as the spindle and spindle fibers. Chromosomes will also be visible under a microscope and will be connected at the centromere. During this condensation and alignment period in meiosis, the homologous chromosomes undergo a break in their double-stranded DNA at the same locations followed by a recombination of the now fragmented parental DNA strands into non-parental combinations, known as crossing over. This process is evidenced to be caused in a large part by the highly conserved Spo11 protein through a mechanism similar to that seen with toposomerase in DNA replication and transcription.

Metaphase

In metaphase, the centromeres of the chromosomes convene themselves on the *metaphase plate* (or *equatorial plate*), an imaginary line that is equidistant from the two centrosome poles and held together by complex complexes known as cohesins. Chromosomes line up in the middle of the cell by microtubule organizing centers (MTOCs) pushing and pulling on centromeres of both chromatids thereby causing the chromosome to move to the center. At this point the chromosomes are still condensing and are currently one step away from being the most coiled and condensed they will be, and the spindle fibers have already connected to the kinetochores. During this phase all the microtubules, with the exception of the kinetochores, are in a state of instability promoting their progression towards anaphase. At this point, the chromosomes are ready to split into opposite poles of the cell towards the spindle to which they are connected.

Anaphase

Anaphase is a very short stage of the cell cycle and occurs after the chromosomes align at the mitotic plate. Kinetochores emit anaphase-inhibition signals until their attachment to the mitotic spindle. Once the final chromosome is properly aligned and attached the final signal dissipates and triggers the abrupt shift to anaphase. This abrupt shift is caused by the activation of the anaphase-promoting complex and its function of tagging degradation of proteins important towards the metaphase-anaphase transition. One of these proteins that is broken down is securin which through its breakdown releases the enzyme separase that cleaves the cohesin rings holding together the sister chromatids thereby leading to the chromosomes separating. After the chromosomes line up in the middle of the cell, the spindle fibers will pull them apart. The chromosomes are split apart as the sister chromatids move to opposite sides of the cell. While the sister chromatids are being pulled apart, the cell and plasma are elongated by non-kinetochore microtubules.

Telophase

Telophase is the last stage of the cell cycle in which a cleavage furrow splits the cells cytoplasm (cytokinesis) and chromatin. This occurs through the synthesis of a new nuclear envelopes that forms around the chromatin gathered at each pole and the reformation of the nucleolus as the chromosomes decondense their chromatin back to the loose state it possessed during interphase. The division of the cellular contents is not always equal and can vary by cell type as seen with oocyte formation where one of the four daughter cells possess the majority of the cytoplasm.

Variants

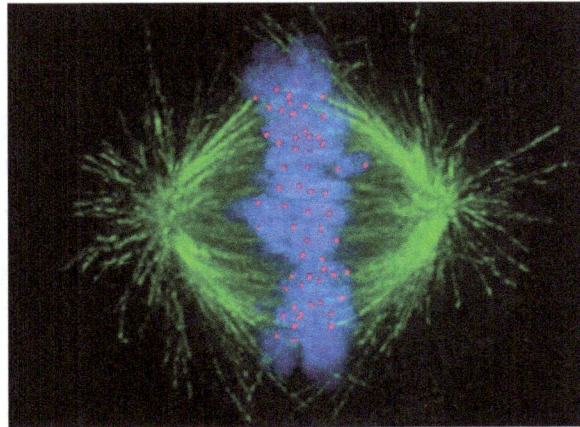

Image of the mitotic spindle in a human cell showing microtubules in green,
chromosomes (DNA) in blue, and kinetochores in red.

Cells are broadly classified into two main categories: simple, non-nucleated prokaryotic cells, and complex, nucleated eukaryotic cells. Due to their structural differences, eukaryotic and prokaryotic cells do not divide in the same way. Also, the pattern of cell division that transforms eukaryotic stem cells into gametes (sperm cells in males or egg cells in females), termed meiosis, is different from that of the division of somatic cells in the body. Image of the mitotic spindle in a human cell showing microtubules in green, chromosomes (DNA) in blue, and kinetochores in red.

Cell division over 42. The cells were directly imaged in the cell culture vessel,
using non-invasive quantitative phase contrast time-lapse microscopy.

Degradation

Multicellular organisms replace worn-out cells through cell division. In some animals, however, cell division eventually halts. In humans this occurs, on average, after 52 divisions, known as the Hayflick limit. The cell is then referred to as senescent. With each division the cells telomeres, protective sequences of DNA on the end of a chromosome that prevent degradation of the chromosomal DNA, shorten. This shortening has been correlated to negative effects such as age related diseases and shortened lifespans in humans. Cancer cells, on the other hand, are not thought to degrade in this way, if at all. An enzyme complex called telomerase, present in large quantities in cancerous cells, rebuilds the telomeres through synthesis of telomeric DNA repeats, allowing division to continue indefinitely.

Cell Signalling

Cell signaling (cell signalling in British English) is part of any communication process that governs basic activities of cells and coordinates multiple-cell actions. The ability of cells to perceive and correctly respond to their microenvironment is the basis of development, tissue repair, and immunity, as well as normal tissue homeostasis. Errors in signaling interactions and cellular information processing may cause diseases such as cancer, autoimmunity, and diabetes. By understanding cell signaling, clinicians may treat diseases more effectively and, theoretically, researchers may develop artificial tissues.

Systems biology studies the underlying structure of cell-signaling networks and how changes in these networks may affect the transmission and flow of information (signal transduction). Such networks are complex systems in their organization and may exhibit a number of emergent properties, including bistability and ultrasensitivity. Analysis of cell-signaling networks requires a combination of experimental and theoretical approaches, including the development and analysis of simulations and modeling. Long-range allostery is often a significant component of cell-signaling events.

All cells receive and respond to signals from their surroundings. This is accomplished by a variety of signal molecules that are secreted or expressed on thw surface of one cell and bine to receptor expressed by the other cells, thereby integrating and coordinating the function of the many individual cells that make up organisms. Each cell is programmed to respond to specific extracellular signal molecules. Extracellularly signaling usually ininvolves the following steps:

- Synthesis and release of the signaling molecule by the signaling cell;

- Transport of the signaling to the target cell;

- Binding of the singnal by a specific receptor leading to its activation;

- Initiation of signal-transduction pathways.

Between Organisms

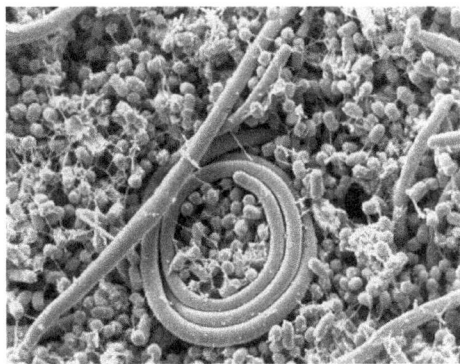

Example of signaling between bacteria. *Salmonella enteritidis* uses N-Acyl homoserine lactone for Quorum sensing.

Cell signaling has been most extensively studied in the context of human diseases and signaling between cells of a single organism. However, cell signaling may also occur between the cells of two different organisms. In many mammals, early embryo cells exchange signals with cells of the uterus. In the human gastrointestinal tract, bacteria exchange signals with each other and with human epithelial and immune system cells. For the yeast *Saccharomyces cerevisiae* during mating, some cells send a peptide signal (mating factor pheromones) into their environment. The mating factor peptide may bind to a cell surface receptor on other yeast cells and induce them to prepare for mating.

Classification

Cell signaling can be classified as either mechanical or biochemical based on the type of the signal. Mechanical signals are the forces exerted on the cell and the forces produced by the cell. These forces can both be sensed and responded to by the cells. Biochemical signals are the biochemical molecules such as proteins, lipids, ions and gases. These signals can be categorized based on the distance between signaling and responder cells. Signaling within, between, and amongst cells is subdivided into the following classifications:

- Intracrine signals are produced by the target cell that stay within the target cell.

- Autocrine signals are produced by the target cell, are secreted, and affect the target cell itself via receptors. Sometimes autocrine cells can target cells close by if they are the same type of cell as the emitting cell. An example of this are immune cells.

- Juxtacrine signals target adjacent (touching) cells. These signals are transmitted along cell membranes via protein or lipid components integral to the membrane and are capable of affecting either the emitting cell or cells immediately adjacent.

- Paracrine signals target cells in the vicinity of the emitting cell. Neurotransmitters represent an example.

- Endocrine signals target distant cells. Endocrine cells produce hormones that travel through the blood to reach all parts of the body.

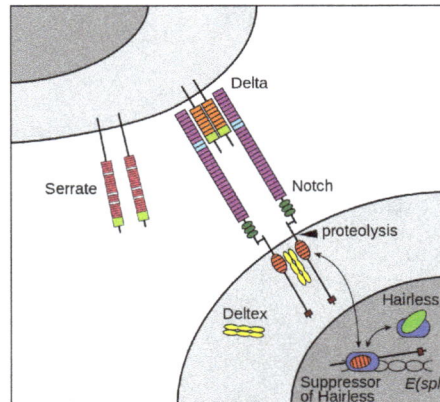

Notch-mediated juxtacrine signal between adjacent cells.

Cells communicate with each other via direct contact (juxtacrine signaling), over short distances (paracrine signaling), or over large distances and/or scales (endocrine signaling).

Some cell–cell communication requires direct cell–cell contact. Some cells can form gap junctions that connect their cytoplasm to the cytoplasm of adjacent cells. In cardiac muscle, gap junctions between adjacent cells allows for action potential propagation from the cardiac pacemaker region of the heart to spread and coordinate contraction of the heart.

The notch signaling mechanism is an example of juxtacrine signaling (also known as contact-dependent signaling) in which two adjacent cells must make physical contact in order to communicate. This requirement for direct contact allows for very precise control of cell differentiation during embryonic development. In the worm *Caenorhabditis elegans,* two cells of the developing gonad each have an equal chance of terminally differentiating or becoming a uterine precursor cell that continues to divide. The choice of which cell continues to divide is controlled by competition of cell surface signals. One cell will happen to produce more of a cell surface protein that activates the Notch receptor on the adjacent cell. This activates a feedback loop or system that reduces Notch expression in the cell that will differentiate and that increases Notch on the surface of the cell that continues as a stem cell.

Many cell signals are carried by molecules that are released by one cell and move to make contact with another cell. *Endocrine* signals are called hormones. Hormones are produced by endocrine cells and they travel through the blood to reach all parts of the body. Specificity of signaling can be controlled if only some cells can respond to a particular hormone. *Paracrine* signals such as retinoic acid target only cells in the vicinity of the emitting cell. Neurotransmitters represent another example of a paracrine signal. Some signaling molecules can function as both a hormone and a neurotransmitter. For example, epinephrine and norepinephrine can function as hormones when released from the adrenal gland and are transported to the heart by way of the blood stream. Norepinephrine can also be produced by neurons to function as a neurotransmitter within the brain. Estrogen can be released by the ovary and function as a hormone or act locally via paracrine or autocrine signaling. Active species of oxygen and nitric oxide can also act as cellular messengers. This process is dubbed redox signaling.

In Multicellular Organisms

In a multicellular organism, signaling between cells occurs either through release into the extracellular space, divided in paracrine signaling (over short distances) and endocrine signaling (over long distances), or by direct contact, known as juxtacrine signaling. Autocrine signaling is a special case of paracrine signaling where the secreting cell has the ability to respond to the secreted signaling molecule. Synaptic signaling is a special case of paracrine signaling (for chemical synapses) or juxtacrine signaling (for electrical synapses) between neurons and target cells. Signaling molecules interact with a target cell as a ligand to cell surface receptors, and/or by entering into the cell through its membrane or endocytosis for intracrine signaling. This generally results in the activation of second messengers, leading to various physiological effects.

A particular molecule is generally used in diverse modes of signaling, and therefore a classification by mode of signaling is not possible. At least three important classes of signaling molecules are widely recognized, although non-exhaustive and with imprecise boundaries, as such membership is non-exclusive and depends on the context:

- Hormones are the major signaling molecules of the endocrine system, though they often regulate each other's secretion via local signaling (e.g. islet of Langerhans cells), and most are also expressed in tissues for local purposes (e.g. angiotensin) or failing that, structurally related molecules are (e.g. PTHrP).

- Neurotransmitters are signaling molecules of the nervous system, also including neuropeptides and neuromodulators. Neurotransmitters like the catecholamines are also secreted by the endocrine system into the systemic circulation.

- Cytokines are signaling molecules of the immune system, with a primary paracrine or juxtacrine role, though they can during significant immune responses have a strong presence in the circulation, with systemic effect (altering iron metabolism or body temperature). Growth factors can be considered as cytokines or a different class.

Signaling molecules can belong to several chemical classes: lipids, phospholipids, amino acids, monoamines, proteins, glycoproteins, or gases. Signaling molecules binding surface receptors are generally large and hydrophilic (e.g. TRH, Vasopressin, Acetylcholine), while those entering the cell are generally small and hydrophobic (e.g. glucocorticoids, thyroid hormones, cholecalciferol, retinoic acid), but important exceptions to both are numerous, and a same molecule can act both via surface receptor or in an intracrine manner to different effects. In intracrine signaling, once inside the cell, a signaling molecule can bind to intracellular receptors, other elements, or stimulate enzyme activity (e.g. gasses). The intracrine action of peptide hormones remains a subject of debate.

Hydrogen sulfide is produced in small amounts by some cells of the human body and has a number of biological signaling functions. Only two other such gases are currently known to act as signaling molecules in the human body: nitric oxide and carbon monoxide.

Signaling Receptors

Cells receive information from their neighbors through a class of proteins known as receptors. Notch is a cell surface protein that functions as a receptor. Animals have a small set of genes that code for signaling proteins that interact specifically with Notch receptors and stimulate a response in cells that express Notch on their surface. Molecules that activate (or, in some cases, inhibit) receptors can be classified as hormones, neurotransmitters, cytokines, and growth factors, in general called receptor ligands. Ligand receptor interactions such as that of the Notch receptor interaction, are known to be the main interactions responsible for cell signaling mechanisms and communication.

The notch acts as a receptor for ligands that are expressed on adjacent cells. While some receptors are cell surface proteins, others are found inside cells. For example, estrogen is a hydrophobic molecule that can pass through the lipid bilayer of the membranes. As part of the endocrine system, intracellular estrogen receptors from a variety of cell types can be activated by estrogen produced in the ovaries.

A number of transmembrane receptors for small molecules and peptide hormones, as well as intracellular receptors for steroid hormones exist, giving cells the ability to respond to a great number of hormonal and pharmacological stimuli. In diseases, often, proteins that interact with receptors are aberrantly activated, resulting in constitutively activated downstream signals.

For several types of intercellular signaling molecules that are unable to permeate the hydrophobic cell membrane due to their hydrophilic nature, the target receptor is expressed on the membrane. When such a signaling molecule activates its receptor, the signal is carried into the cell usually by means of a second messenger such as cAMP.

Signaling Pathways

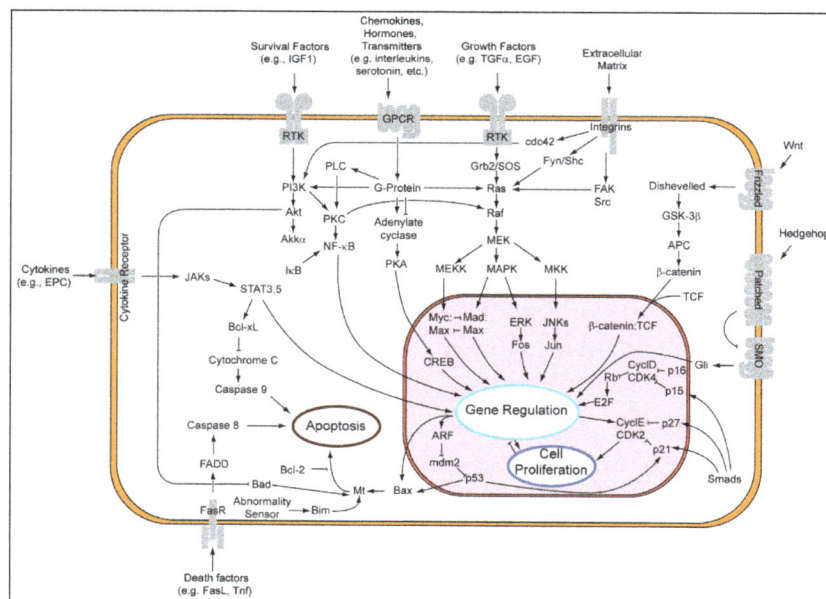

Overview of signal transduction pathways.

In some cases, receptor activation caused by ligand binding to a receptor is directly coupled to the cell's response to the ligand. For example, the neurotransmitter GABA can activate a cell surface receptor that is part of an ion channel. GABA binding to a $GABA_A$ receptor on a neuron opens a chloride-selective ion channel that is part of the receptor. $GABA_A$ receptor activation allows negatively charged chloride ions to move into the neuron, which inhibits the ability of the neuron to produce action potentials. However, for many cell surface receptors, ligand-receptor interactions are not directly linked to the cell's response. The activated receptor must first interact with other proteins inside the cell before the ultimate physiological effect of the ligand on the cell's behavior is produced. Often, the behavior of a chain of several interacting cell proteins is altered following receptor activation. The entire set of cell changes induced by receptor activation is called a signal transduction mechanism or pathway.

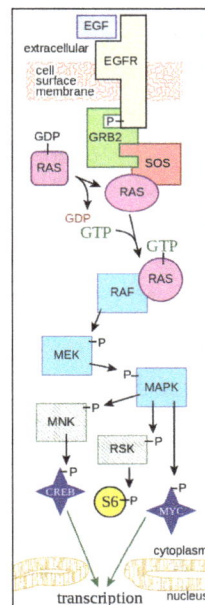

Key components of a signal transduction pathway (MAPK/ERK pathway shown).

In the case of Notch-mediated signaling, the signal transduction mechanism can be relatively simple. As shown in figure, activation of Notch can cause the Notch protein to be altered by a protease. Part of the Notch protein is released from the cell surface membrane and takes part in gene regulation. Cell signaling research involves studying the spatial and temporal dynamics of both receptors and the components of signaling pathways that are activated by receptors in various cell types.

A more complex signal transduction pathway is shown in figure. This pathway involves changes of protein–protein interactions inside the cell, induced by an external signal. Many growth factors bind to receptors at the cell surface and stimulate cells to progress through the cell cycle and divide. Several of these receptors are kinases that start to phosphorylate themselves and other proteins when binding to a ligand. This phosphorylation can generate a binding site for a different protein and thus induce protein–protein interaction. In figure, the ligand (called epidermal growth factor (EGF)) binds to the receptor (called EGFR). This activates the receptor to phosphorylate itself. The phosphorylated receptor binds to an adaptor protein (GRB2), which couples the signal to further downstream signaling processes. For example, one of the signal transduction pathways that are activated is called the mitogen-activated protein kinase (MAPK) pathway. The signal transduction component labeled as "MAPK" in the pathway was originally called "ERK," so the pathway is called the MAPK/ERK pathway. The MAPK protein is an enzyme, a protein kinase that can attach phosphate to target proteins such as the transcription factor MYC and, thus, alter gene transcription and, ultimately, cell cycle progression. Many cellular proteins are activated downstream of the growth factor receptors (such as EGFR) that initiate this signal transduction pathway.

Some signaling transduction pathways respond differently, depending on the amount of signaling received by the cell. For instance, the hedgehog protein activates different genes, depending on the amount of hedgehog protein present.

Complex multi-component signal transduction pathways provide opportunities for feedback, signal amplification, and interactions inside one cell between multiple signals and signaling pathways.

Intra and Inter-species Signaling

Molecular signaling can occur between different organisms, whether unicellular or multicellular. The emitting organism produces the signaling molecule, secretes it into the environment, where it diffuses, and it is sensed or internalized by the receiving organism. In some cases of interspecies signaling, the emitting organism can actually be a host of the receiving organism, or vice versa.

Intraspecies signaling occurs especially in bacteria, yeast, social insects, but also many vertebrates. The signaling molecules used by multicellular organisms are often called pheromones. They can have such purposes as alerting against danger, indicating food supply, or assisting in reproduction. In unicellular organisms such as bacteria, signaling can be used to 'activate' peers from a dormant state, enhance virulence, defend against bacteriophages, etc. In quorum sensing, which is also found in social insects, the multiplicity of individual signals has the potentiality to create a positive feedback loop, generating coordinated response. In this context, the signaling molecules are called autoinducers. This signaling mechanism may have been involved in evolution from unicellular to multicellular organisms. Bacteria also use contact-dependent signaling, notably to limit their growth.

Molecular signaling can also occur between individuals of different species. This has been particularly studied in bacteria. Different bacterial species can coordinate to colonize a host and participate in common quorum sensing. Therapeutic strategies to disrupt this phenomenon are being investigated. Interactions mediated through signaling molecules are also thought to occur between the gut flora and their host, as part of their commensal or symbiotic relationship. Gram negative microbes deploy bacterial outer membrane vesicles for intra- and inter-species signaling in natural environments and at the host-pathogen interface.

Additionally, interspecies signaling occurs between multicellular organisms. In *Vespa mandarinia*, individuals release a scent that directs the colony to a food source.

Computational Models

Recent approaches to better understand elements of pathway crosstalk, complex ligand-receptor binding, and signaling network dynamics have been aided by the use of systems biology approaches. Computational models often take aim at compiling information from published literature to generate a coherent set of signaling components and their associated interactions. The development of computational models allows for a more in-depth probing of cell signaling pathways at a global level by manipulating different variables and systemically evaluating the resulting response. The use of analytical models for the study of signal transduction has been heavily applied in the fields of pharmacology and drug discovery to assess receptor-ligand interactions and pharmacokinetics as well as the flow of metabolites in large networks. A commonly applied strategy to model cell signaling mechanisms is through the use of ordinary differential equation (ODE) models by expressing the time-dependent concentration of a signaling molecule as a function of other molecules downstream and/or upstream within the pathway. ODE models have already been applied for dynamic analysis of the Mitogen-activated protein kinase, Estrogen receptor alpha, and MTOR signaling pathways among numerous others.

Cell Death

Cell death is the event of a biological cell ceasing to carry out its functions. This may be the result of the natural process of old cells dying and being replaced by new ones, or may result from such factors as disease, localized injury, or the death of the organism of which the cells are part. Apoptosis or Type I cell-death, and autophagy or Type II cell-death are both forms of programmed cell death, while necrosis is a non-physiological process that occurs as a result of infection or injury.

Programmed Cell Death

Programmed cell death (or PCD) is cell death mediated by an intracellular program. PCD is carried out in a regulated process, which usually confers advantage during an organism's life-cycle. For example, the differentiation of fingers and toes in a developing human embryo occurs because cells between the fingers apoptose; the result is that the digits are separate. PCD serves fundamental functions during both plant and metazoa (multicellular animals) tissue development.

Apoptosis

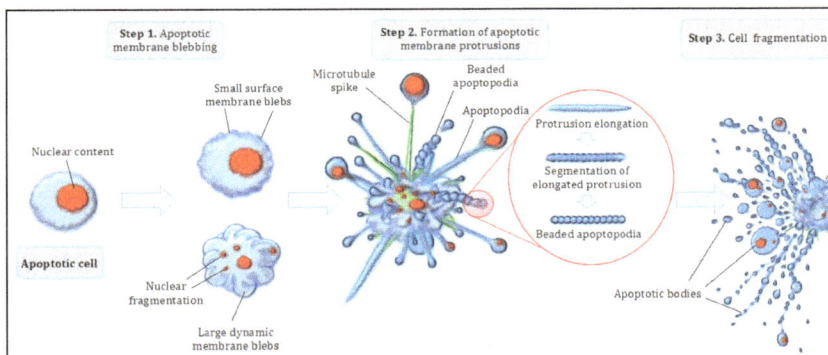

Morphological changes associated with apoptosis.

Apoptosis is the process of programmed cell death (PCD) that may occur in multicellular organisms. Biochemical events lead to characteristic cell changes (morphology) and death. These changes include blebbing, cell shrinkage, nuclear fragmentation, chromatin condensation, and chromosomal DNA fragmentation. It is now thought that – in a developmental context – cells are induced to positively commit suicide whilst in a homeostatic context; the absence of certain survival factors may provide the impetus for suicide. There appears to be some variation in the morphology and indeed the biochemistry of these suicide pathways; some treading the path of "apoptosis", others following a more generalized pathway to deletion, but both usually being genetically and synthetically motivated. There is some evidence that certain symptoms of "apoptosis" such as endonuclease activation can be spuriously induced without engaging a genetic cascade, however, presumably true apoptosis and programmed cell death must be genetically mediated. It is also becoming clear that mitosis and apoptosis are toggled or linked in some way and that the balance achieved depends on signals received from appropriate growth or survival factors.

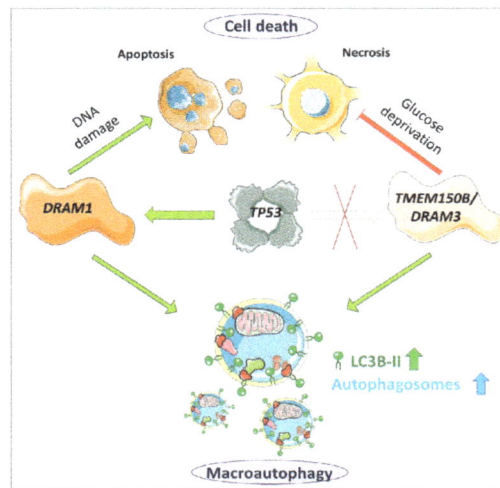

Example events in autophagy.

Autophagy

Autophagy is *cytoplasmic*, characterized by the formation of large vacuoles that eat away organelles in a specific sequence prior to the destruction of the nucleus. Macroautophagy, often referred to as autophagy, is a catabolic process that results in the autophagosomic-lysosomal degradation of bulk cytoplasmic contents, abnormal protein aggregates, and excess or damaged organelles. Autophagy is generally activated by conditions of nutrient deprivation but has also been associated with physiological as well as pathological processes such as development, differentiation, neurodegenerative diseases, stress, infection and cancer.

Other Variations of PCD

Other pathways of programmed cell death have been discovered. Called "non-apoptotic programmed cell-death" (or "caspase-independent programmed cell-death"), these alternative routes to death are as efficient as apoptosis and can function as either backup mechanisms or the main type of PCD.

Some such forms of programmed cell death are anoikis, almost identical to apoptosis except in its induction; cornification, a form of cell death exclusive to the eyes; excitotoxicity; ferroptosis, an iron-dependent form of cell death and Wallerian degeneration.

Plant cells undergo particular processes of PCD similar to autophagic cell death. However, some common features of PCD are highly conserved in both plants and metazoa.

Activation-induced cell death (AICD) is a programmed cell death caused by the interaction of Fas receptor (Fas, CD95)and Fas ligand (FasL, CD95 ligand). It occurs as a result of repeated stimulation of specific T-cell receptors (TCR) and it helps to maintain the periphery immune tolerance. Therefore, an alteration of the process may lead to autoimmune diseases. In the other words AICD is the negative regulator of activated T-lymphocytes.

Ischemic cell death, or oncosis, is a form of accidental, or passive cell death that is often considered a lethal injury. The process is characterized by mitochondrial swelling, cytoplasm vacuolization, and swelling of the nucleus and cytoplasm.

Mitotic catastrophe is a mode of cell death that is due to premature or inappropriate entry of cells into mitosis. It is the most common mode of cell death in cancer cells exposed to ionizing radiation and many other anti-cancer treatments.

Immunogenic cell death or immunogenic apoptosis is a form of cell death caused by some cytostatic agents such as anthracyclines, oxaliplatin and bortezomib, or radiotherapy and photodynamic therapy (PDT).

Pyroptosis is a highly inflammatory form of programmed cell death that occurs most frequently upon infection with intracellular pathogens and is likely to form part of the antimicrobial response in myeloid cells.

Necrotic Cell Death

Necrosis is cell death where a cell has been badly damaged through external forces such as trauma or infection and occurs in several different forms. In necrosis, a cell undergoes swelling, followed by uncontrolled rupture of the cell membrane with cell contents being expelled. These cell contents often then go on to cause inflammation in nearby cells. A form of programmed necrosis, called necroptosis, has been recognized as an alternative form of programmed cell death. It is hypothesized that necroptosis can serve as a cell-death backup to apoptosis when the apoptosis signaling is blocked by endogenous or exogenous factors such as viruses or mutations. Necroptotic pathways are associated with death receptors such as the tumor necrosis factor receptor 1.

Cell Adhesion

All cells rely on cell signaling to detect and respond to cues in their environment. This process not only promotes the proper functioning of individual cells, but it also allows communication and coordination among groups of cells — including the cells that make up organized communities called tissues. Because of cell signaling, tissues have the ability to carry out tasks no single cell could accomplish on its own.

Different types of tissues, such as bone, brain, and the lining of the gut, have characteristic features related to the number and types of cells they contain. Cell spacing is also critical to tissue function, so this geometry is precisely regulated. To preserve proper tissue architecture, adhesive molecules help maintain contact between nearby cells and structures, and tiny tunnel-like junctions allow the passage of ions and small molecules between adjacent cells. Meanwhile, signaling molecules relay positional information among the cells in a tissue, as well as between these cells and the extracellular matrix. These signaling pathways are critical to maintaining the state of equilibrium known as homeostasis within a tissue. For example, the processes involved in wound healing depend on positional information in order for normal tissue architecture to be restored. Such positional signals are also crucial for the development of adult structures in multicellular organisms. As tissues develop, clumps of unorganized cells grow and sort themselves according to signals they send and receive.

Integrins Promote Tissue Structure and Function

Within tissues, adhesive molecules allow cells to maintain contact with one another and with structures in the extracellular matrix. One especially important class of adhesive molecules is the integrins. Integrins are more than just mechanical links, however: They also relay signals both to and from cells. In this way, integrins play an important role in sensing the environment and controlling cell shape and motility.

Integrins are a diverse family of transmembrane proteins found in all animal cells. Even simple animals like sponges have these proteins. Each individual integrin consists of two main parts: an alpha subunit and a beta subunit. Variation in the alpha and beta subunits accounts for the wide variety of integrins observed throughout the animal kingdom. For example, humans alone have over 20 different kinds of integrins.

Integrins link the actin cytoskeleton of a cell to various external structures. The cytoplasmic portion of each integrin molecule binds to adaptor proteins that connect to the actin filaments inside the cell. The extracellular portion of the integrin then binds to molecules in the extracellular matrix or on the surface of other cells. Integrin attachments to neighboring cells can break and reform as a cell moves.

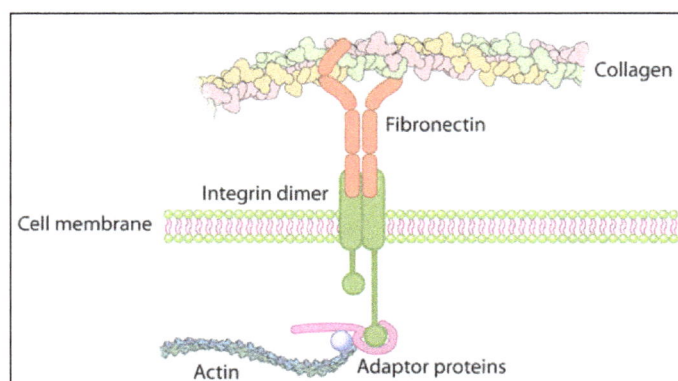

Integrin connects the extracellular matrix with the actin cytoskeleton inside the cell.

How Else do the Cells within a Tissue Stay in Contact?

Beyond integrins, cells rely on several other adhesive proteins to maintain physical contact. As an example, consider the epithelial cells that line the inner and outer surfaces of the human body — including the skin, intestines, airway, and reproductive tract. These cells provide a dramatic example of the different kinds of cell-to-cell junctions, but the same junctions also exist in a wide range of other tissues.

The side surfaces of epithelial cells are tightly linked to those of neighboring cells, forming a sheet that acts as a barrier. Within this sheet, each individual cell has a set orientation. Through integrins, the basal end of each cell connects to a specialized layer of extracellular matrix called the basal lamina. In contrast, the apical end of each cell faces out into the environment — such as the inner cavity or lumen of the gut.

The side-to-side junctions that link epithelial cells are diverse in their protein makeup and function. The adhesive transmembrane proteins anchoring these junctions have extracellular portions that interact with similar proteins on adjacent cells. Protein complexes within each cell further connect the transmembrane adhesive proteins to the cytoskeleton. In particular, adaptor complexes bind adherens junctions to cytoskeletal actin, and other adaptor complexes bind desmosomes to intermediate filaments. Both of these types of junctional complexes provide cells and tissues with mechanical support, and they additionally recruit intracellular signaling molecules to relay positional information to the nucleus.

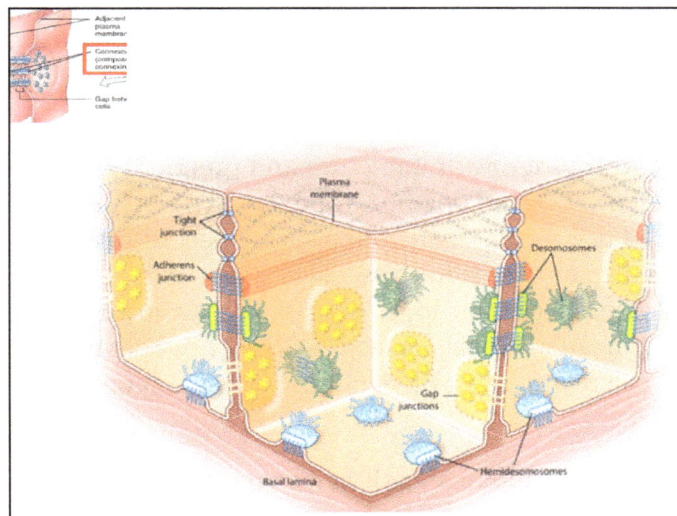

Tight junctions (blue dots) between cells are connected areas of the plasma membrane that stitch cells together. Adherens junctions (red dots) join the actin filaments of neighboring cells together. Desmosomes are even stronger connections that join the intermediate filaments of neighboring cells. Hemidesmosomes (light blue) connect intermediate filaments of a cell to the basal lamina, a combination of extracellular molecules on other cell surfaces. Gap junctions (yellow) are clusters of channels that form tunnels of aqueous connectivity between cells.

The lateral surfaces of epithelial cells also contain several other types of specialized junctions. Tight junctions form a seal between cells that is so strong that not even ions can pass across it. Gap junctions are involved in cellular communication — not just in epithelial tissue, but in other tissue types as well. Gap junctions are specialized connections that form a narrow pore between adjacent cells. These pores permit small molecules and ions to move from one cell to another. In this way, gap junctions provide metabolic and electrical coupling between cells. For example, cardiac tissue has extensive gap junctions, and the rapid movement of ions through these junctions helps the tissue beat in rhythm. Gap junctions may also open and close in response to metabolic signals.

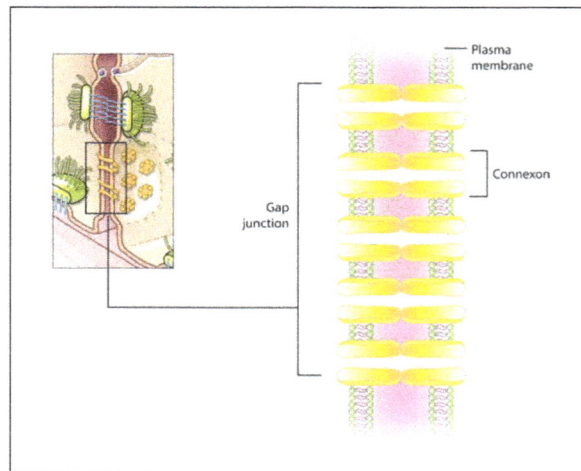

In a gap junction, the lipid bilayer of adjacent cells is pierced through by proteins called connexons. These proteins group together and effectively form a group of communication tunnels between adjacent cells.

Cell Death can be Prompted by a Signal

Cell signaling isn't just central to tissue architecture and function: It also plays an important role in the balance between cell growth and death. Although it sounds like a bad thing, apoptosis — or the process of programmed cell death — is an essential aspect of development. Without it, repair and replenishment processes would overrun tissues with new cells. The orderly demise of a certain proportion of cells is therefore necessary for normal tissue turnover and maintenance of homeostasis. Apoptosis is distinct from necrosis, a messier form of cell death that causes cells to literally swell and burst. Necrotic cell death is not programmed; rather, it occurs in response to trauma or injury.

A range of extracellular and intracellular signals can trigger either cell growth or apoptosis. When cells receive these signals from their neighbors or from other aspects of the external environment, they carefully weigh them against each other before choosing a course of action. For instance, signals that indicate a lack of nutrients or the presence of toxins would likely stall cell growth and promote apoptosis. Within the cell, damage to the DNA or loss of mitochondrial integrity might also result in programmed cell death.

Cells self-destruct cleanly and quickly during apoptosis, thanks to the activation of a variety of enzymes — proteases and nucleases — that break down proteins and nucleic acids, respectively. In fact, scientists look for a characteristic pattern of fragmentation and nuclear condensation within tissues as evidence that apoptosis has occurred.

Some cell signaling occurs on a local level, such as when cells interact with the surrounding extracellular matrix or with their immediate neighbors. This type of signaling is especially important to the structure and function of tissues. Various signaling molecules allow the cells within a tissue to share information about internal and external conditions. This information helps the cells arrange themselves, coordinate their functions, and even know when to grow and when to die. Some of these signaling molecules also function in an adhesive capacity — not just relaying messages between the cells in a tissue, but physically joining these cells to one another.

References

- animal-cells, cells, micro: basicbiology.net, Retrieved 28 April, 2019

- Kang, Joohyun; Park, Jiyoung (December 6, 2011). "Plant ABC Transporters". American Society of Plant Biologists. 9: e0153. doi:10.1199/tab.0153. PMC 3268509. PMID 22303277

- eukaryote, science: britannica.com, Retrieved 16 June , 2019

- Paston, Ira; Willingham, Mark C. (1985). Endocytosis. Springer, Boston, MA. pp 1-44. doi: 10.1007/978-1-4615-6904-6_1. ISBN 9781461569060

- prokaryotes-meaning: thoughtco.com, Retrieved 16 may , 2019

- Sadava, David; H. Craig Heller; Gordon H. Orians; William K. Purves; David M. Hillis (2007). "What are the passive processes of membrane transport?". Life : the science of biology (8th ed.). Sunderland, MA: Sinauer Associates. pp. 105–110. ISBN 9780716776710

- Palmgren, Michael G. (2001-01-01). "PLANT PLASMA MEMBRANE H+-ATPases: Powerhouses for Nutrient Uptake". Annual Review of Plant Physiology and Plant Molecular Biology. 52 (1): 817–845. doi:10.1146/annurev.arplant.52.1.817. PMID 11337417

- Alcamo, I. Edward (1997). "Chapter 2–5: Passive transport". Biology coloring workbook. Illustrations by John Bergdahl. New York: Random House. pp. 24–25. ISBN 9780679778844

- Cells : building blocks of life. Maton, Anthea. (3rd ed.). Upper Saddle River, N.J.: Prentice-Hall. 1997. pp. 70–74. ISBN 978-0134234762. OCLC 37049921

- Ryan RP, Dow JM (July 2008). "Diffusible signals and interspecies communication in bacteria". Microbiology. 154 (Pt 7): 1845–58. doi:10.1099/mic.0.2008/017871-0. PMID 18599814

- cell-adhesion-and-cell-communication, topicpage, scitable: nature.com, Retrieved 21 April, 2019

5
Animal Behavior

Animal behavior, also known as ethology, is a scientific field which studies the behavior of animals under natural conditions. It considers behavior as an evolutionary adaptive trait. The diverse focus areas within the field of animal behavior such as neuroethology and phenotypic plasticity have been thoroughly discussed in this chapter.

Animal behavior is the concept, broadly considered, referring to everything animals do, including movement and other activities and underlying mental processes. Human fascination with animal behavior probably extends back millions of years, perhaps even to times before the ancestors of the species became human in the modern sense. Initially, animals were probably observed for practical reasons because early human survival depended on knowledge of animal behavior. Whether hunting wild game, keeping domesticated animals, or escaping an attacking predator, success required intimate knowledge of an animal's habits. Even today, information about animal behavior is of considerable importance. For example, in Britain, studies on the social organization and the ranging patterns of badgers (Meles meles) have helped reduce the spread of tuberculosis among cattle, and studies of sociality in foxes (Vulpes vulpes) assist in the development of models that predict how quickly rabies would spread should it ever cross the English Channel. Likewise in Sweden, where collisions involving moose (Alces alces) are among the most common traffic accidents in rural areas, research on moose behavior has yielded ways of keeping them off roads and verges. In addition, investigations of the foraging of insect pollinators, such as honeybees, have led to impressive increases in agricultural crop yields throughout the world.

Even if there were no practical benefits to be gained from learning about animal behavior, the subject would still merit exploration. Humans (Homo sapiens) are animals themselves, and most humans are deeply interested in the lives and minds of their fellow humans, their pets, and other creatures. British ethologist Jane Goodall and American field biologist George Schaller, as well as British broadcaster David Attenborough and Australian wildlife conservationist Steve Irwin, have brought the wonders of animal behavior to the attention and appreciation of the general public. Books, television programs, and movies on the subject of animal behavior abound.

Darwin's Influence

The origins of the scientific study of animal behavior lie in the works of various European thinkers of the 17th to 19th centuries, such as British naturalists John Ray and Charles Darwin and French naturalist Charles LeRoy. These individuals appreciated the complexity and apparent purposefulness of the actions of animals, and they knew that understanding behavior demands long-term observations of animals in their natural settings. At first, the principal attraction of natural history studies was to confirm the ingenuity of God. The publication of Darwin's On the Origin of Species in 1859 changed this attitude. In his chapter on instinct, Darwin was concerned with whether behavioral traits, like anatomical ones, can evolve as a result of natural selection. Since then, biologists have recognized that the behaviors of animals, like their anatomical structures, are adaptations that exist because they have, over evolutionary time (that is, throughout the formation of new species and the evolution of their special characteristics), helped their bearers to survive and reproduce.

Charles Darwin.

Furthermore, humans have long appreciated how beautifully and intricately the behaviors of animals are adapted to their surroundings. For example, young birds that possess camouflaged colour patterns for protection against predators will freeze when the parent spots a predator and calls the alarm. Darwin's achievement was to explain how such wondrously adapted creatures could arise from a process other than special creation. He showed that adaptation is an inexorable result of four basic characteristics of living organisms:

1. There is variation among individuals of the same species. Even closely related individuals, such as parent and offspring or sibling and sibling, differ considerably. Familiar human examples include differences in facial features, hair and eye colour, height, and weight.

2. Many of these variations are inheritable—that is, offspring resemble their parents in many traits as a result of the genes they share.

3. There are differences in numbers of surviving offspring among parents in every species. For example, one female snapping turtle (family Chelydridae) may lay 24 eggs; however, only 5 may survive to adulthood. In contrast, another female may lay only 18 eggs, with 1 of her offspring surviving to adulthood.

4. The individuals that are best equipped to survive and reproduce perpetuate the highest frequency of genes to descendant populations. This is the principle known colloquially as "survival of the fittest," where fitness denotes an individual's overall ability to pass copies of his genes on to successive generations. For example, a woman who rears six healthy offspring has greater fitness than one who rears just two.

An inevitable consequence of variation, inheritance, and differential reproduction is that, over time, the frequency of traits that render individuals better able to survive and reproduce in their present environment increases. As a result, descendant generations in a population resemble most closely the members of ancestral populations that were able to reproduce most effectively. This is the process of natural selection.

Function

In studying the function of a behavioral characteristic of an animal, a researcher seeks to understand how natural selection favours the behavior. In other words, the researcher tries to identify the ecological challenges, or "selection pressures," faced by a species and then investigates how a particular behavioral trait helps individuals surmount these obstacles so that they can survive and reproduce.

Until the mid-1960s, functional interpretations of animal behavior were usually made in terms of how a behavior was "good for the species." Social behaviors that excluded some individuals from reproducing (such as territorial defense and courtship displays) were seen as adaptations for regulating animal populations at levels that would prevent overpopulation, environmental destruction, and extinction of the species. This view was based on the observation of ecological phenomena— such as the overgrazing of grassland by cattle, leading to the starvation of the animals. American evolutionary biologist George C. Williams and British ornithologist David Lack, however, revealed the underlying theoretical problem with the view that animals behave in ways that limit their reproduction for the good of their species. Williams noted that individuals who maximize their own reproduction will have greater genetic success than those who behave in ways that limit their reproduction. Thus, over time, in subsequent generations, reproduction-reducing behaviors will be replaced by reproduction-enhancing ones. Therefore, it has become evident that it is incorrect to interpret the behavior of animals as having evolved to function "for the good of the species." Instead, the appropriate interpretation is how a behavior has evolved for the "good of the individual."

Williams's theoretical argument was bolstered by Lack's long-term study of the reproductive behavior of the European, or common, swift (Apus apus). At first glance, swifts appear to voluntarily restrict their own reproduction. When Lack removed the eggs laid each day from a pair's nest he discovered that the female could lay up to 72 or more eggs in a season. Yet, surprisingly, she usually lays just two or three eggs. Are chimney swifts regulating their egg production to avoid overpopulation, or does the number of eggs laid equal the number of young they can successfully rear each year? Lack answered this question by performing the experiment of adding one or two nestlings to the nests of certain pairs so that, instead of the normal two or three young, they would have to rear four or five. He then compared the reproductive success of these pairs to those that were left rearing the normal number. Lack found that the birds with four or five young were less successful (that is, rearing fewer young to fledging) than those in a control group who reared a normal-sized brood. Therefore, chimney swifts, in rearing just two or three offspring, are not withholding reproduction for the good of their species or local population; instead, they are producing as many young as they can successfully rear given a limited food supply, thereby maximizing their own reproduction.

Chimney swifts provide just one example of a pattern that has been found repeatedly by biologists studying the behavior and reproduction of animals. They have found that individuals are "selfish," behaving in ways that benefit their own reproduction regardless of its long-term effect on the survival of their species. Sometimes, however, animals engage in apparent altruism (that is, they exhibit behavior that increases the fitness of other individuals by engaging in activities that decrease their own reproductive success). For example, American zoologist Paul Sherman found that female Belding's ground squirrels (Spermophilus beldingi) give staccato whistles that warn nearby conspecifics of a predator's approach but also attract the predator's attention to the caller. Likewise, worker honeybees (Apis mellifera) perform suicidal attacks on intruders to defend their colony, and female lions (Panthera leo) sometimes nurse cubs that are not their own (although some authorities note that such cubs suckle the lioness when she is asleep).

The key insight to understanding the evolution of such self-sacrificial behavior was provided by British evolutionary biologist William D. Hamilton in the mid-1960s. He argued that natural selection favours genetic success, not reproductive success per se, and that individuals can pass copies of their genes on to future generations. Genes are passed from direct parentage (the rearing of offspring and grand-offspring) and by assisting the reproduction of close relatives (such as nieces and nephews), a concept referred to as "inclusive fitness" or "kin selection."

Hamilton devised a formula—now called Hamilton's rule—that specifies the conditions under which reproductive altruism evolves:

$$r \times B > C$$

where B is the benefit (in number of offspring equivalents) gained by the recipient of the altruism, C is the cost (in number of offspring equivalents) suffered by the donor while undertaking the altruistic behavior, and r is the genetic relatedness of the altruist to the beneficiary. Relatedness is the probability that a gene in the potential altruist is shared by the potential recipient of the altruistic behavior. Altruism can evolve in a population if a potential donor of assistance can more than make up for losing C offspring by adding to the population B offspring bearing a fraction r of its genes. For example, a female lion with a well-nourished cub gains inclusive fitness by nursing a starving cub of a full sister because the benefit to her sister (B = one offspring that would otherwise die) more than compensates for the loss to herself (C = approximately one quarter of an offspring), since the survival probability of her own, non-starving cub is only slightly reduced. Given that the average genetic relatedness (that is, r) between two full sisters is 0.5, then according to Hamilton's rule $(0.5 \times 1) > 0.25$. In essence, genes for altruism spread by promoting aid to copies of themselves.

According to this view, which was popularized by British zoologist Richard Dawkins, the most appropriate way of viewing natural selection is from a gene-selection perspective, as embodied in Hamilton's rule. Genes that are best able to guide the organisms that bear them to propagate successfully will persist and proliferate over generations. Consequently, an explanation of the function of a particular behavior should include how the behavior promotes the success of the genes that underlie the behavior. Of course, since an animal's behavior almost always promotes genetic success by helping the animal survive and reproduce its genes, investigations of behavioral function typically address the survival and reproductive value of the behavior.

Natural Selection in Action

The most straightforward way to study the function of a behavior is to see how natural selection operates on it under current conditions by studying differential reproduction. Often this kind of investigation can be conducted by exploiting the naturally occurring variation among individuals, such as in a particular phenotypic (observable) trait in a population. Sometimes, however, the researcher must experimentally enhance behavioral variation where too little exists in nature. The experimental approach may have the disadvantage of involving unnatural variants, but it has the advantage of revealing how differences among individuals, even in a single trait, can cause variation in reproductive fitness. Either way, a study of natural selection acting on behavior requires that the researcher be able to observe natural populations and obtain detailed information on each individual's survival, its ability to attract a mate, its fertility, and so forth. All of this information is essential to assess an animal's success in passing on its genes.

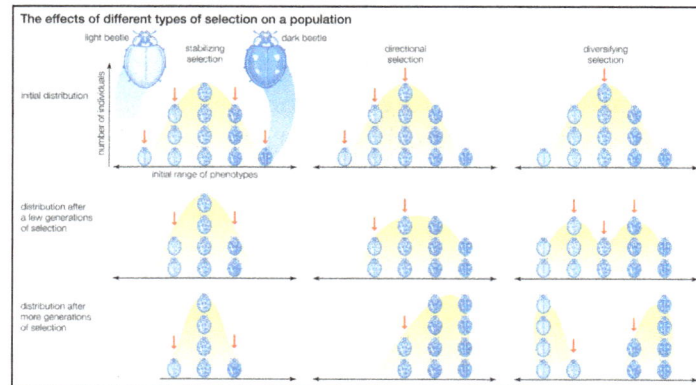

Three types of natural selection, showing the effects of each on the distribution of phenotypes within a population. The downward arrows point to those phenotypes against which selection acts. Stabilizing selection (left column) acts against phenotypes at both extremes of the distribution, favouring the multiplication of intermediate phenotypes. Directional selection (centre column) acts against only one extreme of phenotypes, causing a shift in distribution toward the other extreme. Diversifying selection (right column) acts against intermediate phenotypes, creating a split in distribution toward each extreme.

An investigation of why male titmice, or great tits (Parus major), woodland birds of Europe, sing multiple songs serves to illustrate how a behavioral function can be studied by exploiting naturally existing variation. Each great tit male has a repertoire of one to eight songs that he uses to advertise his presence on a territory. Investigators can acquire detailed information on the breeding biology of these birds because great tits are cavity nesters that readily accept man-made nest boxes. In one experiment on a wooded estate near Oxford, Eng., English zoologist John Krebs and his colleagues installed and regularly inspected nest boxes during the breeding season. The researchers recorded the singing behavior of each breeding male in order to determine repertoire size. They also recorded the egg-laying date, the clutch size (number of eggs), the brood size (number of young), and the fledgling weight for the nests of numerous males. It was possible to monitor the survival of each male's young to the time of its own breeding, because all the young were banded before they fledged and most fledglings returned to the same woods to breed themselves.

The researchers found that individual tits had different repertoire sizes. Males with larger repertoires had chicks that were heavier at fledging, and more of these chicks survived to breed than offspring of males with smaller repertoires. Thus male repertoire size and reproductive success were correlated. The underlying mechanism is that males with larger song repertoires were able to acquire superior territories—specifically, ones with better food. Previous studies had shown that size and survival of young tits depend on body weight at time of fledging: the bigger and heavier the fledgling, the greater its chances of survival to maturity. Thus, the function of a great tit male's singing multiple songs is to help him secure a top-quality breeding territory and mate. So why do all males not sing multiple songs? Perhaps songs are learned over time, so that only the oldest males can possess a large repertoire. Alternatively, perhaps there are costs (such as time away from foraging or increased vulnerability to predators) to singing multiple songs, and only the biggest, strongest males can sing many songs and still survive.

Direct comparisons of individuals of the same species exhibiting natural variation in behavior is a revealing way to study behavioral function. However, when appropriate natural variations do not exist, experimental manipulations can provide the needed variation in the behavior. The variant forms are then studied in the field to determine how well extreme forms of the behavior do in the face of natural selection. Using this method, American biologist Thomas Seeley investigated nest site choice in a species of Southeast Asian honeybee, Apis florea. Colonies build their nests of beeswax combs amid dense foliage, suspended from the branches of bushes and understory trees. Moreover, if a colony's nest loses its cover during the dry season when many trees shed their leaves, the colony will build its new nest in another leafy site. What is the function of this behavior of nesting in dense vegetation? Is it to prevent the nest from overheating under the strong tropical sun, or to conceal the nest from predators, or both?

To test the antipredator hypothesis, pairs of naturally occurring colonies were identified. Within each pair the vegetation around the nest of one colony, which served as the experimental unit, was removed, leaving only enough to provide shade but rendering it conspicuous to predators. The vegetation surrounding the nest of the second colony, which served as the control, was not removed. Measurements of nest site temperatures one day later revealed no significant differences between the two nests. Within one week, however, four of the seven experimental colonies had been discovered and destroyed by predators (probably monkeys and tree shrews) whereas none of the control nests had suffered any damage. Thus, it appears that A. florea colonies choose dense vegetation as nesting sites primarily to conceal their nests from predators.

Another example of a well-controlled field experiment on the function of behavior is Dutch-born British zoologist and ethologist Nikolaas Tinbergen's pioneering study of eggshell removal by black-headed gulls (Larus ridibundus). In a matter of hours after their eggs hatch, they pick up the empty eggshells, fly off, and drop them well away from the nest. Why should a gull engage in this behavior? One hypothesis was that the sharp edges of the shells might injure the chicks, a danger that is well known to poultry breeders. Another hypothesis was that the white insides of broken shells might attract predators, such as crows and herring gulls flying overhead, and so endanger the brood. To test the latter hypothesis, Tinbergen and his colleagues distributed single gull decoy eggs around the dunes where the black-headed gulls nest, and placed broken eggshells near some of the decoy eggs while leaving others isolated. The investigators found that the eggs near broken shells were preyed upon sooner than the isolated, less conspicuous eggs. Evidently, the removal of broken eggshells from the nest by gulls helps to maintain the camouflage of the brood, thereby reducing predation.

Adaptive Design

Many features of animal behavior are so well suited to their function that it is impossible to imagine that they arose by chance. Echolocation by bats, the nest-building skills of weaver birds (family Ploceidae), and the alarm signals of ground squirrels all serve obvious purposes, and the mechanisms that enable them are remarkably similar to what engineers would design to achieve those ends. However, such adaptive behaviors have no divine designer but instead have arisen through the process of natural selection.

Natural selection is an inherently optimizing process: it favours those versions of an organism's traits, including behavioral ones, which best enable the organism to propagate copies of its genes into future generations over alternative versions with lower fitness. Creating a formal optimality model is one way to infer the adaptive "design" or function of a behavior. Using an engineering or economic model to work out the optimal behavioral solution for a given ecological problem is a way of specifying the best design out of a wide range of alternative possibilities. Therefore, if an optimality model embodies an accurate understanding of the function of a behavior, it can predict the form of the behavior that is observed in nature.

One of the attractions of using optimality models to test hypotheses about functional design is that these models yield quantitative predictions that can be easily tested. If a model's predictions regarding the form of a behavior do not match reality, one knows immediately that the hypothesis expressed in the model is false. For example, foraging honeybees often return to the hive with less than a full load of nectar, and biologists initially assumed this was because a bee maximizes its rate of energy delivery to the hive. The fuller the bee, however, the slower she can fly. As a result, the transportation of a full load was assumed to depress a bee's rate of nectar collection. On the other hand, when the bees were trained to forage from an array of artificial flowers in which each flower offered a fixed amount of nectar and the time spent flying between flowers was varied to alter the duration and cost of foraging, the size of the bee's load did not maximize her net rate of energy delivery to the hive. Further analysis revealed that a bee's decision of when she would return to the hive is based on the maximization of foraging efficiency. Evidently, bees behave so as to achieve the highest foraging efficiency rather than the highest food-delivery rate to the hive.

A classic example of application of the optimality approach to understanding the adaptive design of a behavior is a study of copulation time in the yellow dung fly (Scatophaga stercoraria) by British evolutionary biologist Geoffrey A. Parker. Shortly after cow excrement is deposited in a meadow, it is invaded by female dung flies that come to lay their eggs on the dung and by males seeking to mate with the females. Competition among the males for females is fierce. Sometimes one male succeeds in kicking a rival off a female during copulation and mounts her himself. Unfortunately for the first male, this means that some of the female's eggs will be fertilized by the second male. The longer the first male copulates, the more eggs he fertilizes, but the returns for extra copulation time diminish rapidly. How much time should a male spend copulating with a female? Should he copulate for as long as is needed to fertilize all the eggs (about 100 minutes), or should he quit earlier (or permit himself to be displaced) so that he can go search for a new female? Parker hypothesized that a male dung fly chooses a copulation time that maximizes his overall rate of egg fertilizations. He tested his hypothesis using a graphical optimality model.

Before a male dung fly that has just finished copulating with a female can copulate with a new one, he must spend on average 156 minutes searching for her. Once he has found a new female, the proportion of her eggs fertilized by him as a function of copulation time is set by female physiology, and this has been quantified as a curve based on experimentally measured values. The male cannot shorten the time necessary to find a new female or change the fertilization curve, but he can stop copulating at will. The optimal solution, assuming that his decision regarding copulation duration serves to maximize his rate of egg fertilizations, is to copulate for 41 minutes. Because the average observed copulation time, 38 minutes, is quite close to the predicted time of 41 minutes, it is clear that the Darwinian algorithm underlying a male dung fly's copulation behavior serves to maximize his rate of egg fertilizations.

A second way of studying the adaptive design of a behavior is what Darwin called the comparative method, which takes advantage of the thousands of "natural experiments" that have occurred over evolutionary time (that is, throughout the formation of new species and the evolution of their special characteristics). Here again, specific hypotheses regarding how natural selection has shaped a behavior are tested. Rather than simply examining one species, behavioral researchers collect data from a number of species simultaneously. The idea is to compare the degree to which a particular behavior occurs in each species with the degree to which the hypothesized selection pressure is part of the ecology of each species.

Australian zoologist Peter Jarman was one of the first to use the comparative method to study the diversity of mating systems, specifically among various species of African antelope. In some species, such as the dik-dik (Madoqua), individuals are solitary and cryptic; however, during mating season, they form conspicuous monogamous pairs. Others, such as the black wildebeest (Connochaetes taurinus), form enormous herds. During the breeding season, only a few males control sexual access to a group of females in a polygynous mating system. When Jarman compared these African ungulates, he found that body size, typical habitat, group size, and mating system were interrelated. Specifically, smaller species with relatively high metabolic rates (such as the dik-dik) need to consume high-quality food—such as fruits and buds in the forests—while concealing themselves from predators. Because of the sparse distribution of food and the need to remain solitary and cryptic to avoid capture by predators, the smaller species are widely dispersed, leaving no opportunity for a single male to monopolize access to many females. Consequently, small-bodied species tend to be monogamous. In contrast, the larger species graze in open plains where food is generally abundant, although seasonally variable in its geographic distribution, and they are highly visible to predators. Thus, species such as the wildebeest live in large herds that migrate with the seasons. Each individual may be hidden within the large number of other animals in the herd; however, group living creates the opportunity for one male to monopolize several females, and polygyny tends to be found in the large-bodied species. This pattern, which holds true for birds and primates as well as ungulates, supports the hypothesis that the mating system of a species is derived from selection pressures associated with food and predation. Selection pressures determine the spatial distribution of females and thus their defensibility by individual males.

Not all comparative analyses of behavior are so broad. Some focus on just one behavior or a morphological correlate of behavior. Consider the case of sexual dimorphism in body size where the males of some species tend to be considerably larger than the females. It had been hypothesized that size is a key advantage in species where males must fight to defend females from rival males. To test the hypothesis that sexual dimorphism was favoured by natural selection, American evolutionary biologist Richard Alexander and his colleagues compared social structure of the breeding group in primates, ungulates, and pinnipeds with their degree of body-size dimorphism. They reported that body size is similar between males and females in species, including humans, where the breeding group typically consists of one male and one female or a few females. Male body size, however, increases compared with female body size in species that breed in groups made up of multiple males and females, and it is highest in species where a single male defends a large group of females. Evidently, male size in primates is an adaptation related to the intensity of male-male physical competition for females.

Ethology

Ethology is the scientific and objective study of animal behavior, usually with a focus on behavior under natural conditions, and viewing behavior as an evolutionarily adaptive trait. Behaviourism as a term also describes the scientific and objective study of animal behavior, usually referring to measured responses to stimuli or to trained behavioral responses in a laboratory context, without a particular emphasis on evolutionary adaptivity. Throughout history, different naturalists have studied aspects of animal behavior. Ethology has its scientific roots in the work of Charles Darwin and of American and German ornithologists of the late 19th and early 20th century, including Charles O. Whitman, Oskar Heinroth, and Wallace Craig. The modern discipline of ethology is generally considered to have begun during the 1930s with the work of Dutch biologist Nikolaas Tinbergen and of Austrian biologists Konrad Lorenz and Karl von Frisch, the three recipients of the 1973 Nobel Prize in Physiology or Medicine. Ethology combines laboratory and field science, with a strong relation to some other disciplines such as neuroanatomy, ecology, and evolutionary biology. Ethologists typically show interest in a behavioral process rather than in a particular animal group, and often study one type of behavior, such as aggression, in a number of unrelated species.

Ethology is a rapidly growing field. Since the dawn of the 21st century researchers have re-examined and reached new conclusions in many aspects of animal communication, emotions, culture, learning and sexuality that the scientific community long thought it understood. New fields, such as neuroethology, have developed.

Understanding ethology or animal behavior can be important in animal training. Considering the natural behaviors of different species or breeds enables trainers to select the individuals best suited to perform the required task. It also enables trainers to encourage the performance of naturally occurring behaviors and the discontinuance of undesirable behaviors.

Relationship with Comparative Psychology

Comparative psychology also studies animal behavior, but, as opposed to ethology, is construed as a sub-topic of psychology rather than as one of biology. Historically, where comparative psychology has included research on animal behavior in the context of what is known about human psychology, ethology involves research on animal behavior in the context of what is known about animal anatomy, physiology, neurobiology, and phylogenetic history. Furthermore, early comparative psychologists concentrated on the study of learning and tended to research behavior in artificial situations, whereas early ethologists concentrated on behavior in natural situations, tending to describe it as instinctive.

The two approaches are complementary rather than competitive, but they do result in different perspectives, and occasionally conflicts of opinion about matters of substance. In addition, for most of the twentieth century, comparative psychology developed most strongly in North America, while ethology was stronger in Europe. From a practical standpoint, early comparative psychologists concentrated on gaining extensive knowledge of the behavior of very few species. Ethologists were more interested in understanding behavior across a wide range of species to facilitate principled comparisons across taxonomic groups. Ethologists have made much more use of such cross-species comparisons than comparative psychologists have.

Instinct

Kelp gull chicks peck at red spot on mother's beak to stimulate regurgitating reflex.

The Merriam-Webster dictionary defines instinct as "A largely inheritable and unalterable tendency of an organism to make a complex and specific response to environmental stimuli without involving reason".

Fixed Action Patterns

An important development, associated with the name of Konrad Lorenz though probably due more to his teacher, Oskar Heinroth, was the identification of fixed action patterns. Lorenz popularized these as instinctive responses that would occur reliably in the presence of identifiable stimuli called sign stimuli or "releasing stimuli". Fixed action patterns are now considered to be instinctive behavioral sequences that are relatively invariant within the species and that almost inevitably run to completion.

One example of a releaser is the beak movements of many bird species performed by newly hatched chicks, which stimulates the mother to regurgitate food for her offspring. Other examples are the classic studies by Tinbergen on the egg-retrieval behavior and the effects of a "supernormal stimulus" on the behavior of graylag geese.

One investigation of this kind was the study of the waggle dance ("dance language") in bee communication by Karl von Frisch.

Learning

Habituation

Habituation is a simple form of learning and occurs in many animal taxa. It is the process whereby an animal ceases responding to a stimulus. Often, the response is an innate behavior. Essentially, the animal learns not to respond to irrelevant stimuli. For example, prairie dogs (*Cynomys ludovicianus*) give alarm calls when predators approach, causing all individuals in the group to quickly scramble down burrows. When prairie dog towns are located near trails used by humans, giving alarm calls every time a person walks by is expensive in terms of time and energy. Habituation to humans is therefore an important adaptation in this context.

Associative learning

Associative learning in animal behavior is any learning process in which a new response becomes associated with a particular stimulus. The first studies of associative learning were made by Russian physiologist Ivan Pavlov, who observed that dogs trained to associate food with the ringing of a bell would salivate on hearing the bell.

Imprinting

Imprinting in a moose.

Imprinting enables the young to discriminate the members of their own species, vital for reproductive success. This important type of learning only takes place in a very limited period of time. Lorenz observed that the young of birds such as geese and chickens followed their mothers spontaneously from almost the first day after they were hatched, and he discovered that this response could be imitated by an arbitrary stimulus if the eggs were incubated artificially and the stimulus were presented during a critical period that continued for a few days after hatching.

Cultural and Observational Learning

Imitation

Imitation is an advanced behavior whereby an animal observes and exactly replicates the behavior of another. The National Institutes of Health reported that capuchin monkeys preferred the company of researchers who imitated them to that of researchers who did not. The monkeys not only spent more time with their imitators but also preferred to engage in a simple task with them even when provided with the option of performing the same task with a non-imitator. Imitation has been observed in recent research on chimpanzees; not only did these chimps copy the actions of another individual, when given a choice, the chimps preferred to imitate the actions of the higher-ranking elder chimpanzee as opposed to the lower-ranking young chimpanzee.

Stimulus and Local Enhancement

There are various ways animals can learn using observational learning but without the process of imitation. One of these is *stimulus enhancement* in which individuals become interested in an object as the result of observing others interacting with the object. Increased interest in an object can result in object manipulation which allows for new object-related behaviors by trial-and-error learning. Haggerty (1909) devised an experiment in which a monkey climbed up the side of a cage, placed its arm into a wooden chute, and pulled a rope in the chute to release food. Another monkey was provided an opportunity to obtain the food after watching a monkey go through this process on four separate occasions. The monkey performed a different method and finally succeeded after trial-and-error. Another example familiar to some cat and dog owners is the ability of their animals to open doors. The action of humans operating the handle to open the door results in the animals becoming interested in the handle and then by trial-and-error, they learn to operate the handle and open the door.

In local enhancement, a demonstrator attracts an observer's attention to a particular location. Local enhancement has been observed to transmit foraging information among birds, rats and pigs. The stingless bee (*Trigona corvina*) uses local enhancement to locate other members of their colony and food resources.

Social Transmission

A well-documented example of social transmission of a behavior occurred in a group of macaques on Hachijojima Island, Japan. The macaques lived in the inland forest until the 1960s, when a group of researchers started giving them potatoes on the beach: soon, they started venturing onto the beach, picking the potatoes from the sand, and cleaning and eating them. About one year later, an individual was observed bringing a potato to the sea, putting it into the water with one hand, and cleaning it with the other. This behavior was soon expressed by the individuals living in contact with her; when they gave birth, this behavior was also expressed by their young - a form of social transmission.

Teaching

Teaching is a highly specialized aspect of learning in which the "teacher" (demonstrator) adjusts their behavior to increase the probability of the "pupil" (observer) achieving the desired end-result of the behavior. For example, killer whales are known to intentionally beach themselves to catch pinniped prey. Mother killer whales teach their young to catch pinnipeds by pushing them onto the shore and encouraging them to attack the prey. Because the mother killer whale is altering her behavior to help her offspring learn to catch prey, this is evidence of teaching. Teaching is not limited to mammals. Many insects, for example, have been observed demonstrating various forms of teaching to obtain food. Ants, for example, will guide each other to food sources through a process called "tandem running," in which an ant will guide a companion ant to a source of food. It has been suggested that the pupil ant is able to learn this route to obtain food in the future or teach the route to other ants. This behavior of teaching is also exemplified by crows, specifically New Caledonian crows. The adults (whether individual or in families) teach their young adolescent offspring how to construct and utilize tools. For example, *Pandanus* branches are used to extract insects and other larvae from holes within trees.

Mating and the Fight for Supremacy

Individual reproduction is the most important phase in the proliferation of individuals or genes within a species: for this reason, there exist complex mating rituals, which can be very complex even if they are often regarded as fixed action patterns. The stickleback's complex mating ritual, studied by Tinbergen, is regarded as a notable example.

Often in social life, animals fight for the right to reproduce, as well as social supremacy. A common example of fighting for social and sexual supremacy is the so-called pecking order among poultry. Every time a group of poultry cohabitate for a certain time length, they establish a pecking order. In these groups, one chicken dominates the others and can peck without being pecked. A second chicken can peck all the others except the first, and so on. Higher level chickens are easily distinguished by their well-cured aspect, as opposed to lower level chickens. While the pecking order is establishing, frequent and violent fights can happen, but once established, it is broken only when other individuals enter the group, in which case the pecking order re-establishes from scratch.

Living in Groups

Several animal species, including humans, tend to live in groups. Group size is a major aspect of their social environment. Social life is probably a complex and effective survival strategy. It may be regarded as a sort of symbiosis among individuals of the same species: a society is composed of a group of individuals belonging to the same species living within well-defined rules on food management, role assignments and reciprocal dependence.

When biologists interested in evolution theory first started examining social behavior, some apparently unanswerable questions arose, such as how the birth of sterile castes, like in bees, could be explained through an evolving mechanism that emphasizes the reproductive success of as many individuals as possible, or why, amongst animals living in small groups like squirrels, an individual would risk its own life to save the rest of the group. These behaviors may be examples of altruism. Of course, not all behaviors are altruistic, as indicated by the table below. For example, revengeful behavior was at one point claimed to have been observed exclusively in *Homo sapiens*. However, other species have been reported to be vengeful including chimpanzees, as well as anecdotal reports of vengeful camels.

Classification of social behaviors		
Type of behavior	Effect on the donor	Effect on the receiver
Egoistic	Increases fitness	Decreases fitness
Cooperative	Increases fitness	Increases fitness
Altruistic	Decreases fitness	Increases fitness
Revengeful	Decreases fitness	Decreases fitness

Benefits and Costs of Group Living

One advantage of group living can be decreased predation. If the number of predator attacks stays the same despite increasing prey group size, each prey may have a reduced risk of predator attacks through the dilution effect. Further, according to the selfish herd theory theory, the fitness bene-fits associated with group living vary depending on the location of an individual within the group. The theory suggests that conspecifics positioned at the centre of a group will reduce the likelihood predations while those at the periphery will become more vulnerable to attack. Additionally, a predator that is confused by a mass of individuals can find it more difficult to single out one target. For this reason, the zebra's stripes offer not only camouflage in a habitat of tall grasses, but also the advantage of blending into a herd of other zebras. In groups, prey can also actively reduce their predation risk through more effective defence tactics, or through earlier detection of predators through increased vigilance.

Another advantage of group living can be an increased ability to forage for food. Group members may exchange information about food sources between one another, facilitating the process of resource location. Honeybees are a notable example of this, using the waggle dance to communi-cate the location of flowers to the rest of their hive. Predators also receive benefits from hunting in groups, through using better strategies and being able to take down larger prey.

Some disadvantages accompany living in groups. Living in close proximity to other animals can facilitate the transmission of parasites and disease, and groups that are too large may also experi-ence greater competition for resources and mates.

Group Size

Theoretically, social animals should have optimal group sizes that maximize the benefits and min-imize the costs of group living. However, in nature, most groups are stable at slightly larger than optimal sizes. Because it generally benefits an individual to join an optimally-sized group, despite slightly decreasing the advantage for all members, groups may continue to increase in size until it is more advantageous to remain alone than to join an overly full group.

Neuroethology

Neuroethology refers to the study of the neural basis of natural behavior in animals. It attempts to understand how sensory organs and central structures process behaviorally relevant stimuli, and how this information is integrated by the central nervous system to produce the behavioral output observed under natural conditions. Many of the concepts and techniques of neuroethology are derived from other biological disciplines, including ethology, neurophysiology, neuroanatomy, neuroendocrinology, and biological cybernetics. A characteristic overall goal of neuroethology is to understand, from mechanistic and evolutionary points of view, both specialization and diversity of neural control among different species.

Choice of Suitable Model Systems

As in many other biological disciplines, neuroethological research crucially depends on the choice of suitable model systems. Ideally, the behavior under scrutiny should be simple, robust, readily accessible, and ethologically relevant. Thus, such behaviors are exhibited not only under natural conditions, but they can also be evoked without much difficulty, and even on repeated occasions, upon presentation of an adequate stimulus under standardized laboratory conditions. Furthermore, these behavioral patterns are clearly defined so that quantitative analysis is possible. The animal displaying such behaviors should be inexpensive and suitable for maintenance and breeding in the laboratory. Furthermore, the neural network underlying the behavior should be relatively simple in the sense that the nervous system consists of a rather small number of neurons, representing a minimal number of different classes of nerve cells.

Because of the difficulty of obtaining neurophysiological recordings from moving animals, often the only way to monitor the neural activity associated with the perception of sensory stimuli relevant for eliciting a given behavior, or with the generation of the corresponding motor activity, is to employ reduced preparations. Such preparations can be obtained by removing muscles or, most commonly, by immobilizing the animal through blocking synaptic transmission at the neuromuscular junction. Immobilized awake animals can still perceive and process sensory stimuli, and may still be able to generate the neural activity associated with the production of the motor action. These fictive behaviors approximate the real behavior and can be measured with relative ease. For example, tadpoles of the clawed-toad (Xenopus laevis) produce a well-characterized behavior called escape swimming upon sensory stimulation during the first day after hatching. The neural activity of the circuitry in the spinal cord that controls this behavior can be studied in animals that are immobilized by blocking synaptic transmission at the neuromuscular junction with alpha-bungarotoxin (a constituent protein of the venom of the Southeast Asian Krait, Bungarus multicinctus). The study of such fictive swimming in tadpoles has enabled investigators to identify universal mechanisms that control rhythmic motor patterns in vertebrates.

Classical Model Systems in Neuroethology

In contrast to biomedically oriented disciplines, which typically focus on a few model organisms, neuroethology is distinguished by the diversity of taxa studied. The wealth of data obtained through investigations that ask similar questions but are conducted in different species offers the opportunity to examine evolutionary aspects by employing a comparative approach. For example, in gastropod molluscs, similarities and differences in feeding behavior among different species have been linked to anatomical, physiological, and pharmacological properties of the underlying neural network. Neuroethological investigations in African mormyrid fishes have examined the role of sexual signal evolution in relation to morphological and ecological divergence during species radiation. Comparative studies in weakly electric fishes have provided the basis for a phylogenetic comparison of neural systems specialized for time coding.

The following brief description of four classical model systems aims at exemplifying neuroethological research:

- Recognition of prey and predators in toads. This was one of the first major neuroethological research endeavors, led by Jörg-Peter Ewert in the 1960s. It was aimed at the question whether there are neurons that respond selectively to specific features of prey- or predator-like stimuli (so-called feature detectors). By combination of behavioral and physiological experiments, neurons were found in the thalamic-pretectum and the optic tectum, each of which receives visual input from retinal ganglion cells. A specific subpopulation of neurons in the thalamic-pretectum responds best to predator-like features, and electrical stimulation of this area activates escape behavior. On the other hand, a specific subpopulation of neurons in the optic tectum responds best to prey-like features, and stimulation of the brain region evokes prey-catching behavior in toads. Applying a sophisticated method, records of action potentials from freely moving toads showed that the efficiency in prey detection and predator detection are ensured by pretecto-tectal inhibitory interactions. After lesioning pretectal connections to the optic tectum, both prey-selective neurons (recorded by a microelectrode) and prey-catching behavior were disinhibited. After such treatment, these neurons can be activated by any moving object, including predator-like stimuli.

- The neural basis of acoustic communication in crickets. In crickets, communication is largely carried by acoustic signals that are produced by males and transmitted to females or other males. Calling songs, for example, are generated to attract sexually receptive females. A female cricket that is in the state of copulatory readiness responds to these songs by flying or walking toward the source of sound, until she reaches the male (positive phonotaxis). Behavioral experiments have shown that the phonotactic response of female crickets is best elicited by a sound of the carrier frequency and the syllable rate matching those of natural songs produced by the male. This behavioral preference to certain sound parameters is reflected by the tuning to the carrier frequency of the tympanal membrane of the cricket's ear, and an optimal response of certain auditory interneurons to syllable repetition rates in the range that best elicits phonotaxis in females in behavioral tests. These neurons evidently function as recognition neurons.

- The jamming avoidance response of the weakly electric fish of the genus Eigenmannia. A central theme of neuroethology is based on the question of how sensory information is integrated with motor programs to produce a specific, ethologically relevant behavior in response to an adequate stimulus. The first behavior for which a comprehensive answer to this question could be provided was the jamming avoidance response of the weakly electric fish Eigenmannia sp. This behavior consists of shifting the frequency of the fish's electric-organ discharge away from the neighbor's frequency to avoid 'jamming', which would impair the fish's ability to electrolocate objects in its vicinity. Using this model system, it has been possible to identify the entire neural chain underlying this behavior — from the sensory receptors to the effector organ and including the major behavioral and neural rules that govern the sensory processing and the generation of the behavioral output. An important discovery made in the course of the investigations was that different physical parameters of the external electric stimulus triggering the jamming avoidance response are, at lower levels of central processing, analyzed separately (parallel processing), but converge at higher brain levels.

- Neuromodulation of the stomatogastric ganglion of decapod crustaceans. Neural networks are rarely static. Instead, they often exhibit the potential for neural plasticity. This phenomenon forms the basis for behavioral plasticity. Endogenous control of plastic changes in behavior is evident in cases in which stimuli arising from the environment are held constant. The mechanisms that mediate such changes have been studied in great detail in the stomatogastric ganglion of decapod crustaceans. Using this model system, it has been shown that the mode of function of the ganglion is determined by the actual modulatory environment — the anatomical network provides only a physical backbone upon which neuromodulators, particularly neuropeptides and monoamines, can operate. This makes it possible that a single neural network can produce multiple variations in behavioral output under different conditions. For the latter concept, Peter Getting and Michael Dekin coined the term polymorphic network.

Phenotypic Plasticity

Phenotypic plasticity refers to some of the changes in an organism's behavior, morphology and physiology in response to a unique environment. Fundamental to the way in which organisms cope with environmental variation, phenotypic plasticity encompasses all types of environmentally induced changes (e.g. morphological, physiological, behavioral, phenological) that may or may not be permanent throughout an individual's lifespan. The term was originally used to describe developmental effects on morphological characters, but is now more broadly used to describe all phenotypic responses to environmental change, such as acclimation (acclimatization), as well as learning. The special case when differences in environment induce discrete phenotypes is termed polyphenism.

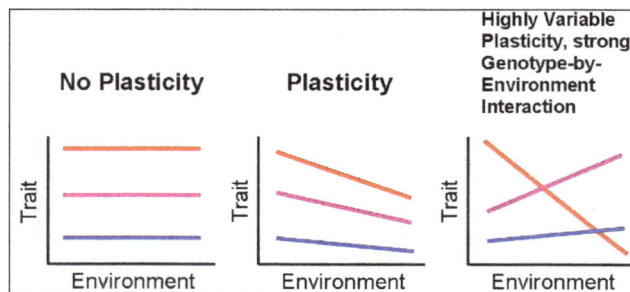

Phenotypic plasticity is the ability of one genotype to produce more than one phenotype when exposed to different environments. Each line here represents a genotype. Horizontal lines show that the phenotype is the same in different environments; slanted lines show that there are different phenotypes in different environments, and thus indicate plasticity.

Generally, phenotypic plasticity is more important for immobile organisms (e.g. plants) than mobile organisms (e.g. most animals), as mobile organisms can often move away from unfavourable environments. Nevertheless, mobile organisms also have at least some degree of plasticity in at least some aspects of the phenotype. One mobile organism with substantial phenotypic plasticity is *Acyrthosiphon pisum* of the aphid family, which exhibits the ability to interchange between asexual and sexual reproduction, as well as growing wings between generations when plants become too populated.

Examples:

*   Animals

The developmental effects of nutrition and temperature have been demonstrated. The gray wolf (*Canis lupus*) has wide phenotypic plasticity. Additionally, male speckled wood butterflies have two morphs: one with three dots on its hindwing, and one with four dots on its hindwings. The development of the fourth dot is dependent on environmental conditions – more specifically, location and the time of year. In amphibians, *Pristimantis mutabilis* has remarkable phenotypic plasticity. Another example is the southern rockhopper penguin. Rockhopper penguins are present at a variety of climates and locations; Amsterdam Island's subtropical waters, Kerguelen Archipelago's subarctic coastal waters, and Crozet Archipelago's subantarctic coastal waters. Due to the species plasticity they are able to express different strategies and foraging behaviors depending on the climate and environment. A main factor that has influenced the species' behavior is where food is located.

- Temperature

Plastic responses to temperature are essential among ectothermic organisms, as all aspects of their physiology are directly dependent on their thermal environment. As such, thermal acclimation entails phenotypic adjustments that are found commonly across taxa, such as changes in the lipid composition of cell membranes. Temperature change influences the fluidity of cell membranes by affecting the motion of the fatty acyl chains of glycerophospholipids. Because maintaining membrane fluidity is critical for cell function, ectotherms adjust the phospholipid composition of their cell membranes such that the strength of van der Waals forces within the membrane is changed, thereby maintaining fluidity across temperatures.

- Diet

Phenotypic plasticity of the digestive system allows some animals to respond to changes in dietary nutrient composition, diet quality, and energy requirements.

Changes in the nutrient composition of the diet (the proportion of lipids, proteins and carbohydrates) may occur during development (e.g. weaning) or with seasonal changes in the abundance of different food types. These diet changes can elicit plasticity in the activity of particular digestive enzymes on the brush border of the small intestine. For example, in the first few days after hatching, nestling house sparrows (*Passer domesticus*) transition from an insect diet, high in protein and lipids, to a seed based diet that contains mostly carbohydrates; this diet change is accompanied by two-fold increase in the activity of the enzyme maltase, which digests carbohydrates. Acclimatizing animals to high protein diets can increase the activity of aminopeptidase-N, which digests proteins.

Poor quality diets (those that contain a large amount of non-digestible material) have lower concentrations of nutrients, so animals must process a greater total volume of poor-quality food to extract the same amount of energy as they would from a high-quality diet. Many species respond to poor quality diets by increasing their food intake, enlarging digestive organs, and increasing the capacity of the digestive tract (e.g. prairie voles, Mongolian gerbils, Japanese quail, wood ducks, mallards). Poor quality diets also result in lower concentrations of nutrients in the lumen of the intestine, which can cause a decrease in the activity of several digestive enzymes.

Animals often consume more food during periods of high energy demand (e.g. lactation or cold exposure in endotherms), this is facilitated by an increase in digestive organ size and capacity, which is similar to the phenotype produced by poor quality diets. During lactation, common degus (*Octodon degus*) increase the mass of their liver, small intestine, large intestine and cecum by 15–35%. Increases in food intake do not cause changes in the activity of digestive enzymes because nutrient concentrations in the intestinal lumen are determined by food quality and remain unaffected. Intermittent feeding also represents a temporal increase in food intake and can induce dramatic changes in the size of the gut; the Burmese python (*Python molurus bivittatus*) can triple the size of its small intestine just a few days after feeding.

AMY2B (Alpha-Amylase 2B) is a gene that codes a protein that assists with the first step in the digestion of dietary starch and glycogen. An expansion of this gene in dogs would enable early dogs to exploit a starch-rich diet as they fed on refuse from agriculture. Data indicated that the wolves and dingo had just two copies of the gene and the Siberian Husky that is associated with hunter-gatherers had just three or four copies, whereas the Saluki that is associated with the Fertile Crescent where agriculture originated had 29 copies. The results show that on average, modern dogs have a high copy number of the gene, whereas wolves and dingoes do not. The high copy number of AMY2B variants likely already existed as a standing variation in early domestic dogs, but expanded more recently with the development of large agriculturally based civilizations.

- Parasitism

Infection with parasites can induce phenotypic plasticity as a means to compensate for the detrimental effects caused by parasitism. Commonly, invertebrates respond to parasitic castration or increased parasite virulence with fecundity compensation in order to increase their reproductive output, or fitness. For example, water fleas (*Daphnia magna*), exposed to microsporidian parasites produce more offspring in the early stages of exposure to compensate for future loss of reproductive success. A reduction in fecundity may also occur as a means of re-directing nutrients to an immune response, or to increase longevity of the host. This particular form of plasticity has been shown in certain cases to be mediated by host-derived molecules (e.g. schistosomin in snails *Lymnaea stagnalis* infected with trematodes *Trichobilharzia ocellata*) that interfere with the action of reproductive hormones on their target organs. Changes in reproductive effort during infection is also thought to be a less costly alternative to mounting resistance or defence against invading parasites, although it can occur in concert with a defence response.

Hosts can also respond to parasitism through plasticity in physiology aside from reproduction. House mice infected with intestinal nematodes experience decreased rates of glucose transport in the intestine. To compensate for this, mice increase the total mass of mucosal cells, cells responsible for glucose transport, in the intestine. This allows infected mice to maintain the same capacity for glucose uptake and body size as uninfected mice.

Phenotypic plasticity can also be observed as changes in behavior. In response to infection, both vertebrates and invertebrates practice self-medication, which can be considered a form of adaptive plasticity. Various species of non-human primates infected with intestinal worms engage in leaf-swallowing, in which they ingest rough, whole leaves that physically dislodge parasites from the intestine. Additionally, the leaves irritate the gastric mucosa, which promotes the secretion of gastric acid and increases gut motility, effectively flushing parasites from the system. The term "self-induced adaptive plasticity" has been used to describe situations in which a behavior under selection causes changes in subordinate traits that in turn enhance the ability of the organism to perform the behavior. For example, birds that engage in altitudinal migration might make "trial runs" lasting a few hours that would induce physiological changes that would improve their ability to function at high altitude.

Woolly bear caterpillars (*Grammia incorrupta*) infected with tachinid flies increase their survival by ingesting plants containing toxins known as pyrrolizidine alkaloids. The physiological basis for this change in behavior is unknown; however, it is possible that, when activated, the immune system sends signals to the taste system that trigger plasticity in feeding responses during infection.

- Reproduction

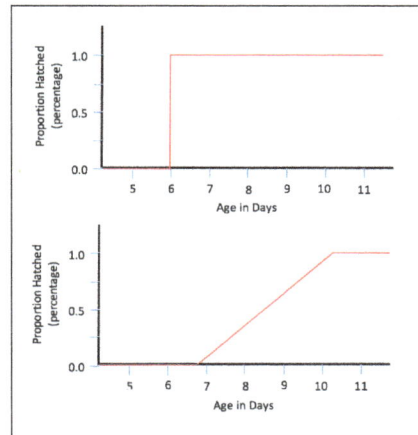

Hatch rates for red-eyed tree frog tadpoles depends on predation

The red-eyed tree frog, *Agalychnis callidryas*, is an arboreal frog (hylid) that resides in the tropics of Central America. Unlike many frogs, the red-eyed tree frog has arboreal eggs which are laid on leaves hanging over ponds or large puddles and, upon hatching, the tadpoles fall into the water below. One of the most common predators encountered by these arboreal eggs is the cat-eyed snake, *Leptodeira septentrionalis*. In order to escape predation, the red-eyed tree frogs have developed a form of adaptive plasticity, which can also be considered phenotypic plasticity, when it comes to hatching age; the clutch is able to hatch prematurely and survive outside of the egg five days after oviposition when faced with an immediate threat of predation. The egg clutches take in important information from the vibrations felt around them and use it to determine whether or not they are at risk of predation. In the event of a snake attack, the clutch identifies the threat by the vibrations given off which, in turn, stimulates hatching almost instantaneously. In a controlled experiment conducted by Karen Warkentin, hatching rate and ages of red-eyed tree frogs were observed in clutches that were and were not attacked by the cat-eyed snake. When a clutch was attacked at six days of age, the entire clutch hatched at the same time, almost instantaneously. However, when a clutch is not presented with the threat of predation, the eggs hatch gradually over time with the first few hatching around seven days after oviposition, and the last of the clutch hatching around day ten. Karen Warkentin's study further explores the benefits and trade-offs of hatching plasticity in the red-eyed tree frog.

Evolution

Plasticity is usually thought to be an evolutionary adaptation to environmental variation that is reasonably predictable and occurs within the lifespan of an individual organism, as it allows individuals to 'fit' their phenotype to different environments. If the optimal phenotype in a given environment changes with environmental conditions, then the ability of individuals to express different traits should be advantageous and thus selected for. Hence, phenotypic plasticity can evolve if Darwinian fitness is increased by changing phenotype. However, the fitness benefits of plasticity can be limited by the energetic costs of plastic responses (e.g. synthesizing new proteins, adjusting expression ratio of isozyme variants, maintaining sensory machinery to detect changes) as well as the predictability and reliability of environmental cues.

Freshwater snails (*Physa virgata*), provide an example of when phenotypic plasticity can be either adaptive or maladaptive. In the presence of a predator, bluegill sunfish, these snails make their shell shape more rotund and reduce growth. This makes them more crush-resistant and better protected from predation. However, these snails cannot tell the difference in chemical cues between the predatory and non-predatory sunfish. Thus, the snails respond inappropriately to non-predatory sunfish by producing an altered shell shape and reducing growth. These changes, in the absence of a predator, make the snails susceptible to other predators and limit fecundity. Therefore, these freshwater snails produce either an adaptive or maladaptive response to the environmental cue depending on whether the predatory sunfish is actually present.

Given the profound ecological importance of temperature and its predictable variability over large spatial and temporal scales, adaptation to thermal variation has been hypothesized to be a key mechanism dictating the capacity of organisms for phenotypic plasticity. The magnitude of thermal variation is thought to be directly proportional to plastic capacity, such that species that have evolved in the warm, constant climate of the tropics have a lower capacity for plasticity compared to those living in variable temperate habitats. Termed the "climatic variability hypothesis", this idea has been supported by several studies of plastic capacity across latitude in both plants and animals. However, recent studies of *Drosophila* species have failed to detect a clear pattern of plasticity over latitudinal gradients, suggesting this hypothesis may not hold true across all taxa or for all traits. Some researchers propose that direct measures of environmental variability, using factors such as precipitation, are better predictors of phenotypic plasticity than latitude alone.

Selection experiments and experimental evolution approaches have shown that plasticity is a trait that can evolve when under direct selection and also as a correlated response to selection on the average values of particular traits.

Group Size Measures

Many animals, including humans, tend to live in groups, herds, flocks, bands, packs, shoals, or colonies (hereafter: groups) of conspecific individuals. The size of these groups, as expressed by the number of people/etc in a group such as eight groups of nine people in each one, is an important aspect of their social environment. Group size tend to be highly variable even within the same species, thus we often need statistical measures to quantify group size and statistical tests to compare these measures between two or more samples. Group size measures are notoriously hard to handle statistically since groups sizes typically follow an aggregated (right-skewed) distribution: most groups are small, few are large, and a very few are very large.

Statistical measures of group size roughly fall into two categories.

Outsiders' view of Group Size

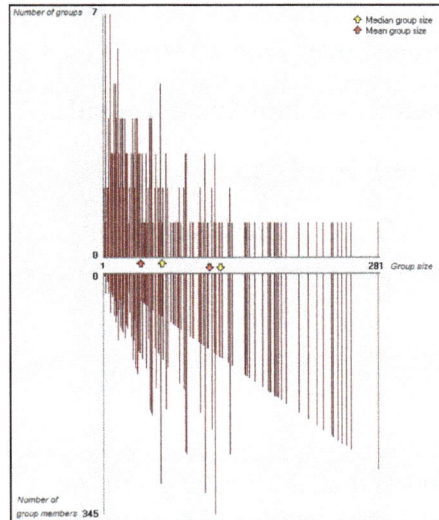

Colony size measures for rooks breeding in Normandy. The distribution of colonies (vertical axis above) and the distribution of individuals (vertical axis below) across the size classes of colonies (horizontal axis). The number of individuals is given in pairs. Animal group size data tend to exhibit aggregated (right-skewed) distributions, i.e. most groups are small, a few are large, and a very few are very large. Note that average individuals live in colonies larger than the average colony size.

- Group size is the number of individuals within a group;

- Mean group size, the arithmetic mean of group sizes averaged over groups;

- Confidence interval for mean group size;

- Median group size, the median of group sizes calculated over groups;

- Confidence interval for median group size.

Insiders' View Of Group Size

As Jarman pointed out, average individuals live in groups larger than average. Therefore, when we wish to characterize a typical (average) individual's social environment, we should apply non-parametric estimations of group size. Reiczigel et al. proposed the following measures:

- Crowding is the size (the number of individuals) of a group that a particular individual lives in (equals to group size: one for a solitary individual, two for both individuals in a group of two, etc.). Practically, it describes the social environment of one particular individual.

- Mean crowding, i.e. the arithmetic mean of crowding measures averaged over individuals (this was called "Typical Group Size" according to Jarman's 1974 terminology);

- Confidence interval for mean crowding.

Example

Imagine a sample with three groups, where group sizes are one, two, and six individuals, respectively, then

Mean group size (group sizes averaged over groups) equals (1 + 2 + 6) / 3 = 3,

Mean crowding (group sizes averaged over individuals) equals,

$$(1+2+2+6+6+6+6+6+6)/9 = 4.555.$$

Generally speaking, given there are G groups with sizes n_1, n_2,..., n_G, mean crowding can be calculated as:

$$\text{Mean crowding} = \sum_{i=1}^{G} n_i^2 / \sum_{i=1}^{G} n_i$$

Statistical Methods

Due to the aggregated (right-skewed) distribution of group members among groups, the application of parametric statistics would be misleading. Another problem arises when analyzing crowding values. Crowding data consist of non-independent values, or ties, which show multiple and simultaneous changes due to a single biological event. (Say, all group members' crowding values change simultaneously whenever an individual joins or leaves.)

Reiczigel et al. discuss the statistical problems associated with group size measures (calculating confidence intervals, two-sample tests, etc.) and offer a free statistical toolset.

Animal Culture

Animal culture involves the current theory of cultural learning in non-human animals, through socially transmitted behaviors. The question as to the existence of culture in non-human societies has been a contentious subject for decades, largely due to the lack of a concise definition for the word "culture". However, many leading scientists agree on seeing culture as a process, rather than an end product. This process, most agree, involves the social transmittance of novel behavior, both among peers and between generations. Such behavior can be shared by a group of animals, but not necessarily between separate groups of the same species.

The notion of culture in animals dates back to Aristotle in classical antiquity, and more recently to Charles Darwin, but the association of animals' actions with the actual word "culture" first originated with Japanese primatologists' discoveries of socially-transmitted food behaviors in the 1940s.

Animal Culture Theory

Though the idea of 'culture' in animals has only been around for just over half of a century, scientists have been noting social behaviors of animals for centuries. Aristotle was the first to provide evidence of social learning in the songs of birds. Charles Darwin first attempted to find the existence of imitation in animals when attempting to prove his theory that the human mind had evolved from that of lower beings. Darwin was also the first to suggest what became known as social learning in attempting to explain the transmission of an adaptive pattern of behavior through a population of honey bees.

The vast majority of cultural anthropological research has been done on non-human primates, due to their being closest evolutionarily to humans. In non-primate animals, research tends to be limited, and therefore evidence for culture strongly lacking. However, the subject has become more popular recently, and has prompted the initiation of more research into the area.

Dawkins's Meme Theory

Evolutionary biologist Richard Dawkins made groundbreaking headway into the field of cultural transmission with his 1976 book entitled *The Selfish Gene*, which focused heavily on the move to evolution being understood primarily by genetic influence. Dawkins coined the term meme, the primary unit of cultural transmission or imitation, to explain an overarching mechanism of how animal behavior is shared and spread to lead to cultural evolution. The use of the word meme was an intentional phonetic derivation of the similar sounding word "gene", which Dawkins asserts to be the primary unit of selection as it lends itself to pathways of biological evolution:

> "We need a name for the new replicator, a noun that conveys the idea of a unit of cultural transmission, or a unit of imitation. 'Mimeme' comes from a suitable Greek root, but I want a monosyllable that sounds a bit like 'gene'. I hope my classicist friends will forgive me if I abbreviate mimeme to meme." (Dawkins)

The analogy between the gene and meme proposed by Dawkins as units of evolutionary biology serves to reinforce the idea that there is a particular pathway of transfer associated with each unit that lends itself either to the evolution of genotypic, phenotypic, and/or behavioral patterns within animal groups. Dawkins asserts that in order for cultural evolution to take place, there needs to be (1) variation within the memes present, (2) the capacity for meme replication between two or more parties, and (3) fitness advantages and/or disadvantages with each meme that lead to the selection or rejection of one meme over another. Likewise, these three criteria in the context of genes are also necessary for genetic evolution. However, with the meme unit, cultural transmission has a distinct feature of being capable of taking place by individuals developing varying interpretations of the meme without exactly "copying" it to pass it on. These interpretations lead to the creation of new memes, which are themselves subject to a cyclic process of selection, rejection, or modification.

Whiten's Culture in Chimpanzees

Andrew Whiten, professor of Evolutionary and Developmental Psychology at the University of St. Andrews, contributed to the greater understanding of cultural transmission with his work on chimpanzees. In *Cultural Traditions in Chimpanzees*, Whiten created a compilation of results from seven long-term studies totaling 151 years of observation analyzing behavioral patterns in different communities of chimpanzees in Africa. The study expanded the notion that cultural behavior lies beyond linguistic mediation, and can be interpreted to include distinctive socially learned behavior such as stone-handling and sweet potato washing in Japanese macaques. The implications of their findings indicate that chimpanzee behavioral patterns mimic the distinct behavioral variants seen in different human populations in which cultural transmission has generally always been an accepted concept.

Cavalli-Sforza and Feldman Models

Population geneticists Cavalli-Sforza & Feldman have also been frontrunners in the field of cultural transmission, describing behavioral "traits" as characteristics pertaining to a culture that are recognizable within that culture. Using a quantifiable approach, Cavalli-Sforza & Feldman were able to produce mathematical models for three forms of cultural transmission, each of which have distinct effects on socialization: vertical, horizontal, and oblique.

- Vertical transmission occurs from parents to offspring and is a function which shows that the probability that parents of specific types give rise to an offspring of their own or of another type. Vertical transmission, in this sense, is similar to genetic transmission in biological evolution as mathematical models for gene transmission account for variation. Vertical transmission also contributes strongly to the buildup of between-population variation.

- Horizontal transmission is cultural transmission taking place among peers in a given population. While horizontal transmission is expected to result in faster within-group evolution due to the relationship building between peers of a population, it is expected to result in less between-group variation than the vertical transmission model would allow for.

- Oblique transmission is cultural transmission being passed from one generation to another younger generation, such as is done by teaching, and the result of reproducing information across generations is a rapid loss of variation within that specific population. Unlike vertical transmission, oblique transmission doesn't need to occur strictly between parent and offspring; it can occur between less-related generations (e.g. from grandparent to grandchild), or from an individual to a non-related younger individual of the same species.

Cultural Transmission in Animals

Cultural transmission, also known as cultural learning, is the process and method of passing on socially learned information. Within a species, cultural transmission is greatly influenced by how adults socialize with each other and with their young. Differences in cultural transmission across species have been thought to be largely affected by external factors, such as the physical environment, that may lead an individual to interpret a traditional concept in a novel way. The environmental stimuli that contribute to this variance can include climate, migration patterns, conflict, suitability for survival, and endemic pathogens. Cultural transmission can also vary according to different social learning strategies employed at the species and or individual level. Cultural transmission is hypothesized to be a critical process for maintaining behavioral characteristics in both humans and nonhuman animals over time, and its existence relies on innovation, imitation, and communication to create and propagate various aspects of animal behavior seen today.

Culture, when defined as the transmission of behaviors from one generation to the next, can be transmitted among animals through various methods. The most common of these methods include imitation, teaching, and language. Imitation has been found to be one of the most prevalent modes of cultural transmission in non-human animals, while teaching and language are much less widespread, with the possible exceptions of primates and cetaceans. Recent research has suggested that teaching, as opposed to imitation, may be a characteristic of certain animals who have more advanced cultural capacities, though this is debatable.

The likelihood of larger groups within a species developing and sharing these intra-species traditions with peers and offspring is much higher than that of one individual spreading some aspect of animal behavior to one or more members. This is why cultural transmission has been shown to be superior to individual learning, as it is a more efficient manner of spreading traditions and allowing members of a species to collectively inherit more adaptive behavior. This process by which offspring within a species acquires his or her own culture through mimicry or being introduced to traditions is referred to as enculturation. The role of cultural transmission in cultural evolution, then, is to provide the outlet for which organisms create and spread traditions that shape patterns of animal behavior visibly over generations.

Genetic vs. Cultural Transmission

Culture, which was once thought of as a uniquely human trait, is now firmly established as a common trait among animals and is not merely a set of related behaviors passed on by genetic transmission as some have argued. Genetic transmission, like cultural transmission, is a means of passing behavioral traits from one individual to another. The main difference is that genetic transmission is the transfer of behavioral traits from one individual to another through genes which are transferred to an organism from its parents during the fertilization of the egg. As can be seen, genetic transmission can only occur once during the lifetime of an organism. Thus, genetic transmission is quite slow compared to the relative speed of cultural transmission. In cultural transmission, behavioral information is passed through means of verbal, visual, or written methods of teaching. Therefore, in cultural transmission, new behaviors can be learned by many organisms in a matter of days and hours rather than the many years of reproduction it would take for a behavior to spread among organisms in genetic transmission.

Social Learning

Culture can be transmitted among animals through various methods, the most common of which include imitation, teaching, and language. Imitation is one of the most prevalent modes of cultural transmission in non-human animals, while teaching and language are much less widespread. In a study on food acquisition techniques in meerkats (*Suricata suricatta*), researchers found evidence that meerkats learned foraging tricks through imitation of conspecifics. The experimental setup consisted of an apparatus containing food with two possible methods that could be used to obtain the food. Naïve meerkats learned and used the method exhibited by the "demonstrator" meerkat trained in one of the two techniques. Although in this case, imitation is not the clear mechanism of learning given that the naïve meerkat could simply have been drawn to certain features of the apparatus from observing the "demonstrator" meerkat and from there discovered the technique on their own.

Teaching

Teaching is often considered one mechanism of social learning, and occurs when knowledgeable individuals of some species have been known to teach others. For this to occur, a teacher must change its behavior when interacting with a naïve individual and incur an initial cost from teaching, while an observer must acquire skills rapidly as a direct consequence.

Until recently, teaching was a skill that was thought to be uniquely human. Now, as research has increased into the transmission of culture in animals, the role of teaching among animal groups has become apparent. Teaching is not merely limited to mammals either. Many insects, for example have been observed demonstrating various forms of teaching in order to obtain food. Ants, for example, will guide each other to food sources through a process called "tandem running", in which an ant will guide a companion ant to a source of food. It has been suggested that the "pupil" ant is able to learn this route in order to obtain food in the future or teach the route to other ants. There have been various recent studies that show that cetaceans are able to transmit culture through teaching as well. Killer whales are known to "intentionally beach" themselves in order to catch and eat pinnipeds who are breeding on the shore. Mother killer whales teach their young to catch pinnipeds by pushing them onto the shore and encouraging them to attack and eat the prey. Because the mother killer whale is altering her behavior in order to help her offspring learn to catch prey, this is evidence of teaching and cultural learning. The intentional beaching of the killer whales, along with other cetacean behaviors such as the variations of songs among humpback whales and the sponging technique used by the bottlenose dolphin to obtain food, provide substantial support for the idea of cetacean cultural transmission.

Teaching is arguably the social learning mechanism that affords the highest fidelity of information transfer between individuals and generations, and allows a direct pathway through which local traditions can be passed down and transmitted.

Imitation

Imitation can be found in a few members of the avian world, in particular the parrot.
Imitation forms the basis of culture, but does not on its own imply culture.

Imitation is often misinterpreted as merely the observation and copying of another's actions. This would be known as mimicry, because the repetition of the observed action is done for no other purpose than to copy the original doer or speaker. In the scientific community, imitation is rather the process in which an organism purposefully observes and copies the methods of another in order to achieve a tangible goal. Therefore, the identification and classification of animal behavior as being imitation has been very difficult. Recent research into imitation in animals has resulted in the tentative labeling of certain species of birds, monkeys, apes, and cetaceans as having the capacity for imitation. For example, a Grey parrot by the name of Alex underwent a series of tests and experiments at the University of Arizona in which scientist Irene Pepperberg judged his ability to imitate the human language in order to create vocalizations and object labels. Through the efforts of Pepperberg, Alex has been able to learn a large vocabulary of English words and phrases. Alex can then combine these words and phrases to make completely new words which are meaningless, but utilize the phonetic rules of the English language. Alex's capabilities of using and understanding more than 80 words, along with his ability to put together short phrases, demonstrates how birds, who many people do not credit with having deep intellect, can actually imitate and use rudimentary language skills in an effective manner. The results of this experiment culminated with the conclusion that the use of the English language to refer to objects is not unique to humans and is arguably true imitation, a basic form of cultural learning found in young children.

Language

Language is another key indicator of animals who have greater potential to possess culture. Though animals do not naturally use words like humans when they are communicating, the well-known parrot Alex demonstrated that even animals with small brains, but are adept at imitation can have a deeper understanding of language after lengthy training. A bonobo named Kanzi has taken the use of the English language even further. Kanzi was taught to recognize words and their associations by using a lexigram board. Through observation of its mother's language training, Kanzi was able to learn how to use the lexigrams to obtain food and other items that he desired. Also, Kanzi is able to use his understanding of lexigrams to decipher and comprehend simple sentences. For example, when he was told to "give the doggie a shot," Kanzi grabbed a toy dog and a syringe and gave it a realistic injection. This type of advanced behavior and comprehension is what scientists have used as evidence for language-based culture in animals.

Primate Culture

A bonobo fishing for termites using a sharpened stick. Tool usage in acquiring food is believed to be a cultural behavior.

The beginning of the modern era of animal culture research in the middle of the 20th century came with the gradual acceptance of the term "culture" in referring to animals. Japan's leading primatologist of the time, Kinji Imanishi, first used the word with a prefix as the term "pre-culture" in referring to the now famous potato-washing behavior of Japanese macaques. In 1948, Imanishi and his colleagues began studying macaques across Japan, and began to notice differences among the different groups of primates, both in social patterns and feeding behavior. In one area, paternal care was the social norm, while this behavior was absent elsewhere. One of the groups commonly dug up and ate the tubers and bulbs of several plants, while monkeys from other groups would not even put these in their mouths. Imanishi had reasoned that, "if one defines culture as learned by offspring from parents, then differences in the way of life of members of the same species belonging to different social groups could be attributed to culture." Following this logic, the differences Imanishi and his colleagues observed among the different groups of macaques may suggest that they had arisen as a part of the groups' unique cultures. The most famous of these eating behaviors was observed on the island of Koshima, where one young female was observed carrying soiled sweet potatoes to a small stream, where she proceeded to wash off all of the sand and dirt before eating. This behavior was then observed in one of the monkey's playmates, then her mother and a few other playmates. The potato-washing eventually spread throughout the whole macaque colony, encouraging Imanishi to refer to the behavior as "pre-culture," explaining that, "we must not overestimate the situation and say that 'monkeys have culture' and then confuse it with human culture." At this point, most of the observed behaviors in animals, like those observed by Imanishi, were related to survival in some way.

A chimpanzee mother and baby.

The first evidence of apparently arbitrary traditions came in the late-1970s, also in the behavior of primates. At this time, researchers McGrew and Tutin found a social grooming handclasp behavior to be prevalent in a certain troop of chimpanzees in Tanzania, but not found in other groups nearby. This grooming behavior involved one chimpanzee taking hold of the hand of another and lifting it into the air, allowing the two to groom each other's armpits. Though this would seem to make grooming of the armpits easier, the behavior actually has no apparent advantage. As the primatologist Frans de Waal explains from his later observations of the hand-clasp grooming behavior in a different group of chimpanzees, "A unique property of the handclasp grooming posture is that it is not required for grooming the armpit of another individual. Thus it appears to yield no obvious benefits or rewards to the groomers."

Prior to these findings, opponents to the idea of animal culture had argued that the behaviors being called cultural were simply behaviors that had evolutionarily evolved due to their importance to survival. After the identification of this initial non-evolutionarily advantageous evidence of culture, scientists began to find differences in group behaviors or traditions in various groups of primates, specifically in Africa. More than 40 different populations of wild chimpanzees have been studied across Africa, between which many species-specific, as well as population-specific, behaviors have been observed. The researching scientists found 65 different categories of behaviors among these various groups of chimpanzees, including the use of leaves, sticks, branches, and stones for communication, play, food gathering or eating, and comfort. Each of the groups used the tools slightly differently, and this usage was passed from chimpanzee to chimpanzee within the group through a complex mix of imitation and social learning.

Chimpanzees

In 1999, Whiten et al. examined data from 151 years of chimpanzee observation in an attempt to discover how much cultural variation existed between populations of the species. The synthesis of their studies consisted of two phases, in which they (1) created a comprehensive list of cultural variant behavior specific to certain populations of chimpanzees and (2) rated the behavior as either customary – occurring in all individuals within that population; habitual – not present in all individuals, but repeated in several individuals; present – neither customary or habitual but clearly identified; absent – instance of behavior not recorded and has no ecological explanation; ecological – absence of behavior can be attributed to ecological features or lack thereof in the environment, or of unknown origin. Their results were extensive: of the 65 categories of behavior studied, 39 (including grooming, tool usage and courtship behaviors) were found to be habitual in some communities but nonexistent in others.

Whiten et al. further made sure that these local traditions were not due to differences in ecology, and defined cultural behaviors as behaviors that are "transmitted repeatedly through social or observational learning to become a population-level characteristic". Eight years later, after "conducting large-scale controlled social-diffusion experiments with captive groups", Whiten et al. stated further that "alternative foraging techniques seeded in different groups of chimpanzees spread differentially across two further groups with substantial fidelity".

This finding confirms not only that nonhuman species can maintain unique cultural traditions; it also shows that they can pass these traditions on from one population to another. The Whiten articles are a tribute to the unique inventiveness of wild chimpanzees, and help prove that humans' impressive capacity for culture and cultural transmission dates back to the now-extinct common ancestor we share with chimpanzees.

Similar to humans, social structure plays an important role in cultural transmission in chimpanzees. Victoria Horner conducted an experiment where an older, higher ranking individual and a younger, lower ranking individual were both taught the same task with only slight aesthetic modification. She found that chimpanzees tended to imitate the behaviors of the older, higher ranking chimpanzee as opposed to the younger, lower ranking individual when given a choice. It is believed that the older higher ranking individual had gained a level of 'prestige' within the group. This research demonstrates that culturally transmitted behaviors are often learned from individuals that are respected by the group.

The older, higher ranking individual's success in similar situations in the past led the other individuals to believe that their fitness would be greater by imitating the actions of the successful individual. This shows that not only are chimpanzees imitating behaviors of other individuals, they are choosing which individuals they should imitate in order to increase their own fitness. This type of behavior is very common in human culture as well. People will seek to imitate the behaviors of an individual that has earned respect through their actions. From this information, it is evident that the cultural transmission system of chimpanzees is more complex than previous research would indicate.

Chimpanzees have been known to use tools for as long as they have been studied. Andrew Whiten found that chimpanzees not only use tools, but also conform to using the same method as the majority of individuals in the group. This conformity bias is prevalent in human culture as well and is commonly referred to as peer pressure.

The results from the research of Victoria Horner and Andrew Whiten show that chimpanzee social structures and human social structures have more similarities than previously thought.

Cetacean Culture

Second only to non-human primates, culture in species within the order Cetacea, which includes whales, dolphins, and porpoises, has been studied for numerous years. In these animals, much of the evidence for culture comes from vocalizations and feeding behaviors.

Cetacean vocalizations have been studied for many years, specifically those of the bottlenose dolphin, humpback whale, killer whale, and sperm whale. Since the early 1970s, scientists have studied these four species in depth, finding potential cultural attributes within group dialects, foraging, and migratory traditions. Hal Whitehead, a leading cetologist, and his colleagues conducted a study in 1992 of sperm whale groups in the South Pacific, finding that groups tended to be clustered based on their vocal dialects. The differences in the whales' songs among and between the various groups could not be explained genetically or ecologically, and thus was attributed to social learning. In mammals such as these sperm whales or bottlenose dolphins, the decision on whether an animal has the capacity for culture comes from more than simple behavioral observations. As described by ecologist Brooke Sergeant, "on the basis of life-history characteristics, social patterns, and ecological environments, bottlenose dolphins have been considered likely candidates for socially learned and cultural behaviors," due to being large-brained and capable of vocal and motor imitation. In dolphins, scientists have focused mostly on foraging and vocal behaviors, though many worry about the fact that social functions for the behaviors have not yet been found. As with primates, many humans are reluctantly willing, yet ever so slightly willing, to accept the notion of cetacean culture, when well evidenced, due to their similarity to humans in having "long lifetimes, advanced cognitive abilities, and prolonged parental care."

Matrilineal Whales

In the cases of three species of matrilineal cetaceans, including pilot whales, sperm whales, and killer whales, mitochondrial DNA nucleotide diversities are about ten times lower than other species of whale. Whitehead found that this low mtDNA nucleotide diversity yet high diversity in matrilineal whale culture may be attributed to cultural transmission, since learned cultural traits have the ability to have the same effect as normal maternally inherited mtDNA. The feeding specializations of these toothed whales are proposed to have led to the divergence of the sympatric "resident" and "transient" forms of killer whales off Vancouver Island, in which resident killer whales feed on fish and squid, and transient whales feed on marine mammals. Vocalizations have also been proven to be culturally acquired in killer and sperm whale populations, as evidenced by the distinct vocalization patterns maintained by members of these different species even in cases where more than one species may occupy one home range. Further study is being done in the matrilineal whales to uncover the cultural transmission mechanisms associated with other advanced techniques, such as migration strategies, new foraging techniques, and babysitting.

Dolphins

By using a "process of elimination" approach, researchers Krutzen et al. reported evidence of culturally transmitted tool use in bottlenose dolphins (*Tursiops* sp.). It has been previously noted that tool use in foraging, called "sponging" exists in this species. "Sponging" describes a behavior where a dolphin will break off a marine sponge, wear it over its rostrum, and use it to probe for fish. Using various genetic techniques, Krutzen et al. showed that the behavior of "sponging" is vertically transmitted from the mother, with most spongers being female. Additionally, they found high levels of genetic relatedness from spongers suggesting recent ancestry and the existence of a phenomenon researchers call a "sponging eve".

In order to make a case for cultural transmission as the mode of behavioral inheritance in this case, Krutzen et al. needed to rule out possible genetic and ecological explanations. Krutzen et al. refer to data that indicate both spongers and nonspongers use the same habitat for foraging. Using mitochondrial DNA data, Krutzen et al. found a significant non-random association between the types of mitochondrial DNA pattern and sponging. Because mitochondrial DNA is inherited maternally, this result suggests sponging is passed from the mother.

In a later study one more possible explanation for the transmission of sponging was ruled out in favor of cultural transmission. Scientists from the same lab looked at the possibility that 1.) the tendency for "sponging" was due to a genetic difference in diving ability and 2.) that these genes were under selection. From a test of 29 spongers and 54 nonspongers, the results showed that the coding mitochondrial genes were not a significant predictor of sponging behavior. Additionally, there was no evidence of selection in the investigated genes.

Rat Culture

Notable research has been done with black rats and Norwegian rats. Among studies of rat culture, the most widely discussed research is that performed by Joseph Terkel in 1991 on a species of black rats that he had originally observed in the wild in Palestine. Terkel conducted an in-depth study aimed to determine whether the observed behavior, the systematic stripping of pine cone scales from pine cones prior to eating, was a socially acquired behavior, as this action had not been observed elsewhere. The experimentation with and observation of these black rats was one of the first to integrate field observations with laboratory experiments to analyze the social learning involved. From the combination of these two types of research, Terkel was able to analyze the mechanisms involved in this social learning to determine that this eating behavior resulted from a combination of ecology and cultural transmission, as the rats could not figure out how to eat the pinecones without being "shown" by mature rats. Though this research is fairly recent, it is often used as a prime example of evidence for culture in non-primate, non-cetacean beings. Animal migration may be in part cultural; released ungulates have to learn over generations the seasonal changes in local vegetation.

In the black rat (*Rattus rattus*), social transmission appears to be the mechanism of how optimal foraging techniques are transmitted. In this habitat, the rats' only source of food is pine seeds that they obtain from pine cones. Terkel et al. studied the way in which the rats obtained the seeds and the method that this strategy was transmitted to subsequent generations. Terkel et al. found that there was an optimal strategy for obtaining the seeds that minimized energy inputs and maximized outputs. Naïve rats that did not use this strategy could not learn it from trial and error or from watching experienced rats. Only young offspring could learn the technique. Additionally, from cross-fostering experiments where pups of naïve mothers were placed with experienced mothers and vice versa, those pups placed with experienced mothers learned the technique while those with naïve mothers did not. This result suggests that this optimal foraging technique is socially rather than genetically transmitted.

Avian Culture

The songs of starlings have been discovered to show regional "dialects," a trait that has potential to have a cultural basis.

Birds have been a strong study subject on the topic of culture due to their observed vocal "dialects" similar to those studied in the cetaceans. These dialects were first discovered by zoologist Peter Marler, who noted the geographic variation in the songs of various songbirds. Many scientists have found that, in attempting to study these animals, they approach a stumbling block in that it is difficult to understand these animals' societies due to their being so different from our own. This makes it difficult to understand the animals' behaviors, let alone determine whether they are cultural or simply practical.

However, despite this hindrance, evidence for differing dialects among songbird populations has been discovered, especially in sparrows, starlings, and cowbirds. In these birds, scientists have found strong evidence for imitation-based learning, one of the main types of social learning. Though the songbirds obviously learn their songs through imitating other birds, many scientists remain skeptical about the correlation between this and culture: "the ability to imitate sound may be as reflexive and cognitively uncomplicated as the ability to breathe. It is how imitation affects and is affected by context, by ongoing social behavior, that must be studied before assuming its explanatory power." The scientists have found that simple imitation does not itself lay the ground for culture, whether in humans or birds, but rather it is how this imitation affects the social life of an individual that matters.

In an experiment looking at vocal behavior in birds, researchers Marler & Tamura found evidence of song dialects in a sparrow species. Located in the eastern and southern parts of North America, *Zonotrichia leucophrys* is a species of white-crowned song-birds that exhibit learned vocal behavior. Marler & Tamura found that while song variation existed between individual birds, each population of birds had a distinct song pattern that varied in accordance to geographical location. For this reason, Marler & Tamura called the patterns of each region a "dialect".

By raising male sparrows in various types of acoustic settings and observing the effects of their verbal behavior, Marler & Tamura found that sparrows learned song "dialects" in about the first 100 days of life from older males. In this experimental setting, male birds in acoustic chambers were exposed to recorded sounds played through a loudspeaker. Using this setup in the laboratory, Marler & Tamura also saw that alien dialects could be taught, and that after learning the dialect, the sparrow's verbal behavior were unaffected by the additional acoustic experiences, like the exposure of other dialects and songs from different species. It was also shown that white-crowned sparrows selectively learn songs of conspecifics. Marler & Tamura note that this case of cultural transmission is interesting in that it requires no social bond between the learner and the emitter of sound, given all emitted sounds originated from a loudspeaker in their experiments.

However, social bond strongly facilitates song imitation in some songbirds. Zebra finches hardly imitate songs played from a loudspeaker, but they can imitate songs of an adult bird after only a few hours of interaction. Interestingly, imitation is inhibited when the number of siblings (pupils) increases. Thus, the imitation process cannot be explained as simple mimicry that depends on the abundance of the model, but is a dynamic process that depends on complex behavioral context.

Observations and Studies of Learned Feeding Behavior in Passerines

In the British Isles, in the 20th century, bottled milk was delivered to households in the early morning (by "milkmen") and left on doorsteps to be collected. Birds, particularly but not exclusively tits (Paridae) and principally blue tit Cyanistes caeruleus, great tit Parus major, coal tit Periparus ater regularly attacked the bottles, opening the foil or cardboard lids, and drank the cream of the top. This behavior was studied, and strong evidence found that learning was involved, although possibly at the level that milk bottles represent a food source, rather than particular techniques. Later studies built on these observations

Fish

Evidence for cultural transmission has also been shown in wild fish populations. Scientists Helfman and Schultz conducted translocation experiments with French grunts (*Haemulon flavolineatum*) where they took fish native to a specific schooling site and transported them to other sites. In this species of fish, the organism uses distinct, traditional migration routes to travel to schooling sites on coral reefs. These routes persisted past one generation and so by relocating the fish to different sites, Helfman and Schultz wanted to see if the new fish could relearn that sites' migration route from the resident fish. Indeed this is what they found: that the newcomers quickly learned the traditional routes and schooling sites. But when residents were removed under similar situations, the new fish did not use the traditional route and instead use new routes, suggesting that the behavior could not be transmitted once the opportunity for learning was no longer there.

Guppy mating behavior is believed to be culturally influenced.

In a similar experiment looking at mating sites in blueheaded wrasse (*Thalassoma bifasciatum*), researcher Warner found that individuals chose mating sites based on social traditions and not based on the resource quality of the site. Warner found that although mating sites were maintained for four generations, when entire local populations were translocated elsewhere, new sites were used and maintained.

Social Learning in Animals

Many of the biologically important decisions that an animal must make can be affected by its observation of the behavior of others. The foods an individual chooses to eat, the motor patterns it uses to gain access to food, the time it spends foraging in a patch, the predators it avoids, the individuals it selects as sex partners—all can be affected by observation of others of its species. Although theory suggests that adopting the behavior of others may not always be in the best interests of an animal observing the actions of conspecifics, the benefits of social learning, particularly to naive young animals, should most often outweigh any potential costs.

Many of the things that young animals have to learn must be learned rapidly. A fledging bird or weaning mammal, venturing from the site where it has been protected and sustained by adults of its species, has to learn to avoid predators before it is eaten by one. It has to learn to select a nutritionally balanced diet before it exhausts its internal reserves of any critical nutrient, and without ingesting harmful quantities of toxins. A naive young animal, newly recruited to a population and faced with such challenges, would be well advised to take advantage of opportunities provided by interaction with adults of its species.

Almost by definition, adults are individuals that have acquired patterns of behavior allowing them to avoid predators and the ingestion of toxins, to select an adequate diet, and to find water and safe refuges. Most important, adults are doing all of these things in the environment where the juveniles with which they interact are struggling to achieve independence. Consequently, to the extent that ignorant juveniles can use the behavior of adults to guide development of their own behavioral repertoires, juveniles should often be able to acquire necessary responses to the demands of the particular locales in which they are living without incurring many of the costs associated with individual trial-and-error learning.

References

- Animal-behavior, science: britannica.com, Retrieved 13 July, 2019

- McGreevy, Paul; Boakes, Robert (2011). Carrots and Sticks: Principles of Animal Training. Darlington Press. pp. xi–23. ISBN 978-1-921364-15-0. Retrieved 9 September 2016

- Jones, Nick A. R.; Rendell, Luke (2018). "Cultural Transmission". Encyclopedia of Animal Cognition and Behavior. Springer, Cham. pp. 1–9. doi:10.1007/978-3-319-47829-6_1885-1. ISBN 978-3-319-47829-6

- Imitation Promotes Social Bonding in Primates, August 13, 2009 News Release". National Institutes of Health. 13 August 2009. Archived from the original on 22 August 2009. Retrieved 8 November2011

- McGreevy, Paul; Boakes, Robert (2011). Carrots and Sticks: Principles of Animal Training. Darlington Press. pp. xi–23. ISBN 978-1-921364-15-0. Retrieved 9 September 2016

- Neuroethology: scholarpedia.org, Retrieved 29 April 2019

- Silvertown, Jonathan (1989). "The paradox of seed size and adaptation". Trends in Ecology & Evolution. 4 (1): 24–26. doi:10.1016/0169-5347(89)90013-x. PMID 21227308

- Bresmer, Brett R.; et al. (7 September 2018). "Is ungulate migration culturally transmitted? Evidence of social learning from translocated animals". Science. 361 (6406): 1023–1025. doi:10.1126/science.aat0985

Permissions

All chapters in this book are published with permission under the Creative Commons Attribution Share Alike License or equivalent. Every chapter published in this book has been scrutinized by our experts. Their significance has been extensively debated. The topics covered herein carry significant information for a comprehensive understanding. They may even be implemented as practical applications or may be referred to as a beginning point for further studies.

We would like to thank the editorial team for lending their expertise to make the book truly unique. They have played a crucial role in the development of this book. Without their invaluable contributions this book wouldn't have been possible. They have made vital efforts to compile up to date information on the varied aspects of this subject to make this book a valuable addition to the collection of many professionals and students.

This book was conceptualized with the vision of imparting up-to-date and integrated information in this field. To ensure the same, a matchless editorial board was set up. Every individual on the board went through rigorous rounds of assessment to prove their worth. After which they invested a large part of their time researching and compiling the most relevant data for our readers.

The editorial board has been involved in producing this book since its inception. They have spent rigorous hours researching and exploring the diverse topics which have resulted in the successful publishing of this book. They have passed on their knowledge of decades through this book. To expedite this challenging task, the publisher supported the team at every step. A small team of assistant editors was also appointed to further simplify the editing procedure and attain best results for the readers.

Apart from the editorial board, the designing team has also invested a significant amount of their time in understanding the subject and creating the most relevant covers. They scrutinized every image to scout for the most suitable representation of the subject and create an appropriate cover for the book.

The publishing team has been an ardent support to the editorial, designing and production team. Their endless efforts to recruit the best for this project, has resulted in the accomplishment of this book. They are a veteran in the field of academics and their pool of knowledge is as vast as their experience in printing. Their expertise and guidance has proved useful at every step. Their uncompromising quality standards have made this book an exceptional effort. Their encouragement from time to time has been an inspiration for everyone.

The publisher and the editorial board hope that this book will prove to be a valuable piece of knowledge for students, practitioners and scholars across the globe.

Index